SCIENTIFIC EXPLANATION AND METHODOLOGY OF SCIENCE

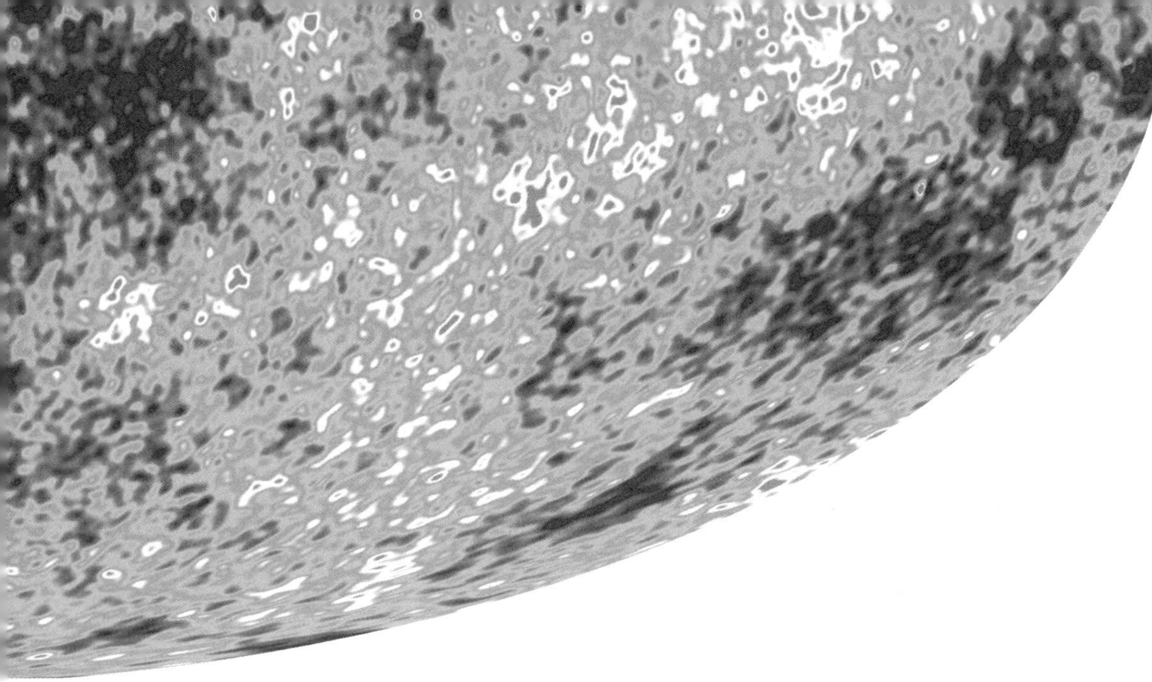

SCIENTIFIC EXPLANATION AND METHODOLOGY OF SCIENCE

Selected Papers from the International Conference on SEMS 2012

Shanxi University, Taiyuan, China, 17–19 September 2012

Editors

Guichun Guo

Shanxi University, China

Chuang Liu

Shanxi University, China & University of Florida, USA

World Scientific

NEW JERSEY • LONDON • SINGAPORE • BEIJING • SHANGHAI • HONG KONG • TAIPEI • CHENNAI

Published by

World Scientific Publishing Co. Pte. Ltd.
5 Toh Tuck Link, Singapore 596224
USA office: 27 Warren Street, Suite 401-402, Hackensack, NJ 07601
UK office: 57 Shelton Street, Covent Garden, London WC2H 9HE

British Library Cataloguing-in-Publication Data
A catalogue record for this book is available from the British Library.

SCIENTIFIC EXPLANATION AND METHODOLOGY OF SCIENCE

ISBN 978-981-4596-63-3

Printed in Singapore by B & Jo Enterprise Pte Ltd

PREFACE

This volume contains the collection of selected papers from The International Conference on Scientific Explanation and Methodology of Science (SEMS 2012), held in September 17-19, 2012, in Taiyuan, Shanxi Province, China. They are revised and augmented versions of the talks presented at the Conference. SEMS 2012 is the first large-scale invitation-only international conference in philosophy of science in China. It included invited international researchers from Asia (besides China), Australia, Europe, Latin America, North America, as well as researchers from within China. Thirty-two scholars in various areas of philosophy of science give talks in the two-day conference program; and from these talks, eighteen are selected for publication in this volume. The papers represent the latest work of each researcher in his or her expertise; and as a result, this volume gives a good representation of the cutting-edge researches in diverse areas in different parts of the world, thereby achieving one of the main goals of the conference.

The conference is made possible by the generous support of the Research Centre for Philosophy of Science and Technology at Shanxi University. We would like to thank the local organizing committee for their tireless work in making sure that every aspect of the meeting is taken good care of. We would also like to thank all the invitees for their enthusiastic acceptance of our invitations and active participation in the program. We thank all the contributors of this volume for taking the time to revise or re-write their papers for the Proceedings. And finally, we want to thank the World Scientific Publishing Company for agreeing to publish the Proceedings and for supplying the copy-editing service.

Guichun Guo
Chuang Liu

ORGANIZATION

Co-Chairmen

Guichun Guo	Shanxi University, China
Chuang Liu	University of Florida, US

Academic Committee

Chair:	Dachun Liu	Renmin University of China, China
Deputy Chairs:	Guichun Guo	Shanxi University, China
	Chuang Liu	University of Florida, USA

Members:

Andreas Bartels	University of Bonn, Germany
James McAllister	University of Leiden, Netherland
James R. Brown	Toronto University, Canada
Elena Castellani	University of Florence, Italy
Anjan Chakravartty	University of Notre Dame, USA
Barry Loewer	Rutgers University, USA
Alex Rosenberg	Duke University, USA
Malcolm Forster	University of Wisconsin- Madison, USA
Fan Chen	Northeastern University, China
Xinhe Hu	Graduate University of Chinese Academy of Sciences, China
Dingcheng Ren	Graduate University of Chinese Academy of Sciences, China
Feng Lu	Tsinghua University, China
Xianjing Xiao	Graduate University of Chinese Academy of Sciences, China
Ce Gao	Shanxi University, China
Jie Yin	Shanxi University, China
Ruijing Qiao	Shanxi University, China
Yidong Wei	Shanxi University, China
Peifu Zhang	Shanxi University, China

The International Conference on Scientific Explanation and Methodology of Science
Shanxi University, Taiyuan, P. R. China 2012. 9

Hosts with guests from the HPS of Notre Dame.
From left: Jun An, Anjan Chakravartty, Chuang Liu, Katherine Brading, Guichun Guo, Janet Kourany, Jie Yin.

List of Participants

	NAME	COLLEGE	COUNTRY/REGION	E-MAIL
1	Guichun Guo	Shanxi University	China	guogc@public.ty.sx.cn
2	Chuang Liu	University of Florida & Shanxi University	USA	logics@ufl.edu
3	Alex Rosenberg	Duke University	USA	alexrose@duke.edu
4	Andreas Bartels	University of Bonn	Germany	andreas.bartels@uni-bonn.de
5	Katherine Brading	University of Notre Dame	USA	kbrading@nd.edu
6	James Robert Brown	Toronto University	Canada	jrbrown@chass.utoronto.ca
7	Anjan Chakravartty	University of Notre Dame	USA	chakravartty.1@nd.edu
8	Kaiyuan Cheng	National Yang-Ming University	Taiwan	kcheng1970@gmail.com
9	Sungho Choi	Kyung Hee University	South Korea	plxsc506@gmail.com
10	Roman Frigg	LSE-London	UK	R.P.Frigg@lse.ac.uk
11	Paul Griffiths	University of Sydney	Australia	paul.griffiths@sydney.edu.au
12	Malcolm Forster	University of Wisconsin-Madison	USA	mforster@wisc.edu
13	Janet A. Kourany	University of Notre Dame	USA	JKourany@nd.edu
14	Barry Loewer	Rutgers University	USA	loewer@rci.rutgers.edu
15	Sergio Martinez	The National Autonomous University of Mexico	Mexico	tepeytualco@gmail.com
16	J. W. McAllister	University of Leiden	Netherlands	j.w.mcallister@phil.leidenuniv.nl

	NAME	COLLEGE	COUNTRY/REGION	E-MAIL
17	Ilkka Niiniluoto	University of Helsinki	Finland	niiniluo@mappi.helsinki.fi
18	Kathleen Okruhlik	University of West Ontario	Canada	okruhlik@uwo.ca
19	Michael Strevens	New York University	USA	strevens@nyu.edu
20	Kelly Trogdon	Lingnan University	Hong Kong	old.jabo@gmail.com
21	Jiji Zhang	Lingnan University	Hong Kong	jijizhang@ln.edu.hk
22	Mary Leng	University of York	Canada	mary.leng@york.ac.uk
23	Wei Wang	Tsinghua University	China	wangwei@tsinghua.edu.cn
24	Xiang Huang	Fudan University	China	hangxiang@fudan.edu.cn
25	Yingjin Xu	Fudan University	China	yjxu@fudan.edu.cn;
26	Xiao Tan	Sun Yat-Sen University	China	tanxiao0706@gmail.com
27	Qiaoying Lu	Sun Yat-Sen University	China	
28	Xianjing Xiao	Chinese Academy of Sciences	China	Xxj201@126.com
29	Liuxiang Hao	Chinese Academy of Sciences	China	
30	Zhu Xu	Chinese Academy of Sciences	China	dennis.zhxu@gucas.ac.cn;
31	Lian Chen	Wuhan University	China	liancheng@whu.edu.cn
32	Hong Fang Li	WuHan University	China	lihongfang@whu.edu.cn
33	Xiaolong Wan	Huazhong University of Science and Technology	China	hwanxl@163.com

	NAME	COLLEGE	COUNTRY/REGION	E-MAIL
34	Gang Chen	Huazhong University of Science and Technology	China	mgchen@hust.edu.cn
35	Mingyi Chen	Huazhong University of Science and Technology	China	cmy_0223@163.com
36	Hong Li	Beijing Normal University	China	lihong01@263.net
37	Guolin Wu	Huanan University of Technology	China	ssglwu@scut.edu.cn
38	Pei Zhang	Tianjin University of Finance and Economis	China	lizzhangpei@163.com
39	Jitang Li	Soochow University	China	lijitang67@163.com
40	Ge Qian	Shanghai University of Finance and Economics	China	qiange@mail.shufe.edu.cn
41	Junli Guo	Shanxi University of Finance and Economics	China	
42	Xiao-hong Wang	Xian Jiaotong University	China	amandawxh@sina.com
43	Ce Gao	Shanxi University	China	rcpst@sxu.edu.cn
44	Jie Yin	Shanxi University	China	yinjie@sxu.edu.cn
45	Yidong Wei	Shanxi University	China	weiyidong@sxu.edu.cn
46	Ruijin Qiao	Shanxi University	China	qrj@sxu.edu.cn
47	Peifu Zhang	Shanxi University	China	zpeifu@sxu.edu.cn
48	Tianping He	Shanxi University	China	htpsx@sxu.edu.cn
49	Jun An	Shanxi University	China	anjun@sxu.edu.cn
50	Shuyan Wang	Shanxi University	China	wshuyan@sxu.edu.cn

	NAME	COLLEGE	COUNTRY/REGION	E-MAIL
51	Shuxue Li	Shanxi University	China	lishuxue@sxu.edu.cn
52	Jie Liu	Shanxi University	China	liujie@sxu.edu.cn
53	Rui Cheng	Shanxi University	China	chengrui@sxu.edu.cn
54	Dan Zhao	Shanxi University	China	chyuzhao@sxu.edu.cn
55	Xiaohua Feng	Shanxi University	China	kinggirlss2001@sohu.com
56	Bin Zhao	Shanxi University	China	sxbin-zhao@163.com
57	Kaining Wang	Shanxi University	China	wangkn@sxu.edu.cn
58	Yang You	Shanxi University	China	youyang@sxu.edu.cn
59	Yaxin Yao	Shanxi University	China	yyx0145@sina.com
60	Jianbo Guo	Shanxi University	China	gojbo@sxu.edu.cn
61	Hongwu Zheng	Shanxi University	China	hwzheng@sxu.edu.cn
62	Lizhen Sun	Shanxi University	China	slz@sxu.edu.cn
63	Yujun Xu	Shanxi University	China	bianjibu@sxu.edu.cn
64	Jie Yu	Shanxi University	China	bianjibu@sxu.edu.cn
65	Dawei Qi	Shanxi University	China	qdw@sxu.edu.cn

CONTENTS

THE DIRECTEDNESS OF TIME IN CLASSICAL COSMOLOGY

ANDREAS BARTELS* & DANIEL WOHLFARTH

Institut für Philosophie, Universität Bonn,
Am Hof 1, 53113 Bonn, Germany
**andreas.bartels@uni-bonn.de*

The aim of this paper is to show that a new understanding of fundamentality can be applied successfully in classical cosmology based on General Relativity. We are thereby able to achieve an account of cosmological time asymmetry as an intrinsic and fundamental property of the universe. First, we consider Price's arguments *against* the fundamental status of time-asymmetry (Price (1996, 2002, 2011)). We show that these arguments have some force, but their force depends on understanding fundamentality as law-likeness. Second, we show that alternative approaches attempting to explain time directedness either by applying an anthropic strategy based on a multiverse approach, or by using the empirical fact of accelerated expansion of the universe, equally fail to provide a fundamental explanation of time directedness. In the third part, we present our own new concept of fundamentality based on properties of the solution space of fundamental laws. We demonstrate how this new concept of fundamentality is effective in understanding the cosmological asymmetry.

Keywords: Time, Cosmology, General Relativity.

1. Searching for a Fundamental Explanation of the Directionality of Time

This paper attempts to show that the directionality of time for cosmological solutions of Einstein's field equations can be explained as a fundamental property of nature. In order to proceed, we first define what it means that a spacetime of General Relativity is time-asymmetric (and thus has the property that its cosmic time is *directed*): That a spacetime having a cosmic time (and therefore having spacelike hypersurfaces) is *time-symmetric* with respect to a hypersurface $\mathrm{t} = \mathrm{t_S}$, intuitively means that, from the hypersurface $\mathrm{t} = \mathrm{t_S}$, the "spacetime looks the same in both temporal directions. Therefore, if a time-orientable spacetime having cosmic time is time-asymmetric, we shall not find a spacelike hypersurface $\mathrm{t} = \mathrm{t_S}$ which splits the spacetime in two 'halves', one the temporal image of the other one with respect to their intrinsic geometrical properties."[1] On the other hand, if we find such a hypersurface, we shall call the spacetime time-symmetric.

[1]cf. Castagnino *et al.* (2009), p. 14. In more technical terms, time-symmetry of a spacetime with respect to a spacelike hypersurface $\mathrm{t} = \mathrm{t_S}$ requires "time-isotropy", i.e. the existence of a diffeomorphism onto itself which reverses the temporal orientations but preserves the metric and leaves the hypersurface $\mathrm{t} = \mathrm{t_S}$ fixed.

An asymmetric spacetime is *time-directed* with respect to the cosmic time parameter in the sense that the two different directions defined by increasing or by decreasing values of the cosmic time parameter can be distinguished by *physical* means. The directedness of time by itself does not lead to an 'arrow of time'. We cannot speak of an 'arrow of time' before we have established connections to other parts of physics, apart from cosmology, such that one of the distinct cosmological time directions can be transferred to some local time direction with respect to which physical systems behave in some distinctive (e.g. entropy increasing) way. In this paper, we will only be concerned with the cosmological directedness of time, not with any 'arrow of time'.

When will we say that the cosmological directedness of time is *fundamentally anchored*? The answer is that in order to be fundamentally anchored, it should be based on *fundamental physical properties* as represented by the theory involved, in our case General Relativity (GR). This does not necessarily imply that the directedness of time is anchored in the fundamental *laws* of the theory, i.e. in the field equations, the laws being only *one* (even if the most general way) of constraining those fundamental properties. In other words: The fundamentality of time directedness does not require the time reversal invariance of fundamental laws.

1.1. *The anti-fundamentality view*

Price starts his cosmological analysis with the observation that the early state of the universe is special in one interesting way: the universe near the Big Bang is smooth. A matter distribution consisting of black holes would be much more probable than a smooth distribution if classical gravity is the dominant force, as assumed in classical cosmology. Thus, according to classical thermodynamics, the 'early' universe has very low entropy. Now, Price (1996, p. 78) argues that entropic behavior by itself cannot provide a directedness of time in classical cosmology: statistical considerations grounding the second law of thermodynamics are also valid in the *reverse* time order. Thus, the time asymmetry of entropic behavior can only be based on the initial low-entropy conditions of the big bang (see also Albert (2000)).

But the attempt to ground time directedness in the specialness of the big bang tends to "fall for the temporal double standard".[2] What makes the very early past so special as compared to the very late future? Imagining the possible case of a re-contracting universe, we can ask what the future final state (the big crunch) would be like: "Would it be a kind of mirror image of the big bang, at least in the sense that it too would display the same remarkable smoothness?"[3] Since our physical theories are time-symmetric and thus "inherently insensitive to temporal direction",[4] it would be "hard to see" how they could imply that one end of the

[2]Price 1996, 80.
[3]Price 1996, 81.
[4]Price 1996, 80.

universe has to be 'special' in this way without implying that both ends have to be 'special'.[5] Thus the answer that a time-symmetric theory urges on us is a positive one: entropy has to be low at *both* ends of the re-contracting universe; it has to be the sort of universe that Gold has described in the early 1960s, with increasing entropy in the expansion phase and symmetrically decreasing entropy in the contracting phase. But if this sort of universe is at least possible, then it does not matter, for the conceptual analysis of time-directedness, if open universes are also possible or even more probably representing our actual universe by empirical reasons: "This point [the possibility of open spacetime geometry, A.B. & D.W.] is an interesting one, but it should not be overrated. For one thing, if we are interested in whether the Gold universe [a kind of a time-symmetric closed spacetime, A.B. & D.W.] is a coherent possibility, the issue as to whether the actual universe recollapses is rather peripheral. [...] Of course, if we could show that a recollapsing universe is impossible, given the laws of physics as we know them, the situation would be rather different".[6]

The upshot of Price's argument is that the mere theoretical possibility of a Gold universe shows that the empirically confirmed past low entropy state of our actual universe cannot provide any clue about time-directedness, to the extent that we stick to the symmetric standard theory (and do not, for instance, resort to some anthropic strategy[7]). The attempt to base time directedness in some past low entropy state must fail because the time-symmetric theory requires treating both of the temporal developments, the transition from now toward the past, and the transition toward the future in the contracting phase of a re-collapsing universe, just alike. Thus, if the one transition leads to a 'past' low entropy state, then its mirror image must lead to some qualitatively identical 'future' low entropy state. Since that latter outcome is in conflict with the natural 'clumping' tendency of gravitating matter,[8] a low entropy big bang must be theoretically puzzling as well and can in any case not be seen as providing a theoretical understanding of time directedness. The situation is even worse: If we could explain the occurrence of a low entropy state at one end of the universe by means of any time-symmetric theory, then we would immediately be faced with the dilemma that this explanation would now apply to the *other* end of the universe as well: no time-asymmetry in, no time-asymmetry out! Price's argument, for sure, is not meant to rule out the *fact* of time-directedness. What is ruled out, according to Price's view, is that time-directedness is *law-governed*, at least with respect to time-symmetric laws.

Now, Price's argument presupposes that 'past' and 'future' low entropy states in a re-contracting universe cannot be distinguished by any theoretical means. But this turns out to imply that the scaling factor which is a symmetrical function of cosmic

[5]Price 1996, 81.
[6]Price 1996, 95.
[7]Cf. Price 1996, 97f.
[8]Cf. Price 1996, 96.

time in a re-contracting universe is the *only* fundamental property that could possibly distinguish past and future time direction. If there were further properties with respect to which the cosmological past could be distinguished from the cosmological future, then Price's argument would not go through.

In fact, there are such additional properties, as will be shown in Section 3. Thus, Price's argument against the fundamentality of time-directedness partly relies on the false assumption that the theory is insensitive to temporal direction. This insensitivity only appears as a result of ignoring physical properties characterizing the models which make those models in fact sensitive to temporal directedness. Thus, it turns out that the models have the potential to favor a particular time direction that the symmetric laws do not have. That 'symmetric' (time reversal invariant) laws may have asymmetric models has also been observed by Price, but he has failed to make use of this fact.

1.2. *The anthropic strategy: Asymmetry explained?*

We now follow a path that, according to Price, could provide a possible way out of the dilemma characterized by the slogan "*no asymmetry in, no asymmetry out*": the *anthropic strategy.* A particular way of realizing this strategy is the *multiverse* account proposed, for instance, by Ćirković & Miloševic-Zdjelar (2004). This account attempts to explain the fact that our universe has a very smooth matter distribution near the big bang (cf. Linde (1990)). Multiverse accounts assume the existence of more than one universe, each of which is characterized by different initial conditions. One of these universes (or: cosmic domains) represents our actual universe. If smooth states near the big bang have a probability to occur of, for example, $1 : 10^{10^{123}}$,[9] the existence of cosmic domains including such smooth early states can be predicted, if the number of cosmic domains is assumed to be much larger than the reciprocal of the probability of the occurrence of such smooth early states. The prediction is of the same kind as the prediction of getting a '6' at least once if a fair six-sided dice is rolled n times, where n is much larger than six. But why do we not observe one of the much more probable domains in our actual cosmic environment? The answer to the question is that no human being could survive in such domains. Here we seem to have an account explaining the occurrence of a low entropy state at one end of the universe, and moreover explaining it in a fundamental way in the sense that basic properties of nature, according to that account, produce the explanation.

But the expectation, according to which the anthropic strategy provides us with a fundamental explanation of temporal asymmetry, is deceptive. The multiverse approach predicts a large number of cosmic domains. In addition, the time parameter, which is fundamental in this context, is a *quantum parameter* existing independently of any particular cosmic times within the particular cosmic domains. The laws that give rise to the fundamental processes by which the variety of cosmic

[9]Cf. Penrose 1979.

domains with their particular cosmic times is created, describe quantum processes, which are time- symmetric according to the quantum physical time parameter.[10] The cosmic domains themselves may be time- symmetric spacetimes with respect to their particular cosmic times. That means that entropy behaves in a symmetric way with respect to past or future development in such spacetimes. Since neither the processes producing the multiverse nor the totality of cosmic domains that is produced by those processes favor any temporal direction, the slogan 'no asymmetry in, no asymmetry out' remains to be valid. Since the anthropic strategy does not adduce any essentially asymmetric theoretical structure, we should not expect that, as a result of that, some asymmetric (temporal) structure occurs.

The anthropic strategy provides us with an explanation for the occurrence of time-asymmetric behaviour in our particular cosmic domain, but this asymmetry is not based on fundamental asymmetric structures. It shows that it should not be surprising to find ourselves in a cosmic environment such as the asymmetric observable universe; nonetheless, the temporal direction is not based on fundamental structures of the theory which is relevant for the strategy, namely quantum field theory.

1.3. *Accelerated expansion: Does it show the fundamentality of time-asymmetry?*

Recent observations and theoretical results indicate that the three-dimensional universe exhibits accelerated expansion in cosmic time, which can be described in General Relativity as a large positive value of the cosmological constant (e.g. Riess *et al.* (1998) on supernovae observations; Barlett and Blanchard (1996) on the cosmic virial theorem; Fan, Bahcall & Cen (1997) on mass indicators in galaxy clusters; Bertschinger (1998) on large-scale velocity maps; Kochanek (1995) and Coles & Ellis (1994)). This new empirical situation that has come to the fore in the years since the publication of Price's book in 1996 seems to indicate that the universe probably has an open geometry, not only because the matter and energy density are too low to overcome expansion but also because the expansion is accelerated by a 'force' described by the cosmological constant.

Now, the question arises whether this large value of the cosmological constant occurs accidentally or according to some law of physics. If the accelerated expansion of the universe were the result of a fundamental law, time asymmetry would thereby get a fundamental status. Ćirković & Miloševic-Zdjelar (2004) have argued that the cosmological constant is a result of vacuum polarisation. Since vacuum polarisation can be derived from the laws of Quantum Field Theory (QFT), it appears as if the cosmological constant could indeed get a lawful status. In the following, we shall show that the above interpretation of the cosmological constant does not lead to

[10] This is true if we assume the standard interpretation of quantum physics, according to which, for example, complex conjugation will not change the physical meaning of a quantum dynamic equation such as the Schrödinger equation.

the conclusion that the open geometry of spacetime, and thus the asymmetry of spacetime are based on laws.

The first reason to deny such a lawful status is that the force of gravity, which opposes the cosmological constant, depends on the matter and energy density of the universe. Thus, the critical value that the cosmological constant must exceed to yield an accelerated universe in such a semi-classical model depends on the matter and energy density. Since this density does not seem to be determined by QFT in general, it remains to be a matter of contingent and empirical fact whether spacetime is open, and thus time-asymmetric.

Second, in the context of multiverses, there is some evidence that this density could vary among different cosmic domains. If such variation is possible, it follows that the effect of the accelerated expansion of the universe could only lead to time asymmetry in some particular cosmic domains.

In general, we think that the new empirical situation does not have an impact on the conceptual situation concerning the question of time asymmetry: a closed universe is still possible even if the value of the cosmological constant is assumed to be 'large' for fundamental reasons. Since time asymmetry has some fundamental status only if the directedness of time is based on fundamental properties of the theory, the approach of Ćirković & Miloševic-Zdjelar (2004) does not provide any new insight concerning this question.

In sum, we have shown that the approaches we discussed in Sections 1.2., 1.3., have turned out to be insufficient in order to support the lawful status of time-asymmetry of the universe. In Section 2 we shall discuss an alternative view, according to which fundamentality is not to be identified with *lawlikeness*. Instead we propose a concept of fundamentality that is based on the behavior of the *solutions* (models) of a physical theory: If a property can be shown to appear by theoretical reasons in *almost all models* of a theory, this property can thereby be classified as a fundamental property. Time-asymmetry with respect to cosmic time, as we shall show, is a *typical* property of a relevant sub-class of solutions of Einstein's field equations of GR. We take the notion of typicality to characterize the behavior of members of a set that can be shown "to hold with very rare exception".[11] In our case of interest, *most* members of a relevant class of solutions of GR field equations manifest asymmetric behavior, i.e. the asymmetric members of the class build a sub-class of *measure 1* with respect to the Lebesque measure on the solution space with respect to natural co-ordinates provided by the scaling factor, the scalar matter field and their respective derivatives.

[11]Cf. Goldstein (2011), p. 2. Typical behavior is manifested, for example, by the motion of atoms of two metal rods at different temperatures that are brought into thermal contact: the motions evolve so that the temperatures in the rods equalize (cf. Maudlin (2007), 287). The notion seems first to appear in Boltzmann's writings. In Boltzmann (1877) he argued that the evolution of a gas in line with the Boltzmann equation applies to the overwhelming majority of phase points (according to any given distribution function) — they evolve into an equilibrium state, that is, the behavior of a gas, while not inevitable, is typical (see: Goldstein (2011), p. 5).

It is worth mentioning that the idea that the *realizations* of a symmetric theory may be exclusively or dominantly *asymmetric* has also been noticed by Price (1996), in the context of his discussion of Stephen Hawking's 'no boundary condition': "[a] symmetric physical theory might be such that all or most of its possible realizations were individually asymmetric. [...] a time-symmetric physical theory might have the consequence that any individual universe has to be asymmetric in time."[12] But Price has remained skeptic as to whether this principal possibility could be used for a solution of the asymmetry problem; thus he claims that "if the required asymmetry were not already implicit in the theoretical premises in some sense, it could not be derived from them".[13]

Thus it appears that, according to Price, fundamentality based on the behavior of (almost all) solutions, has no *autonomous* status. Instead there has to be some derivation from "principles which are not *manifestly* [but in some sense 'implicitly', A.B. & D.W.] asymmetric".[14] Contrary to that, we propose an autonomous kind of fundamentality that is based on the behavior of almost all solutions. What Price did not realize is that the notion of symmetry for *theories* (or laws) does not coincide with the notion of symmetry for their *solutions* (spacetimes in the case of General Relativity). The former means the time reversal invariance of fundamental equations, whereas the latter requires, in the case of General Relativity, that a spacetime, seen from a given space-like hypersurface, looks the same in both temporal directions with respect to cosmic time.

2. A New Concept of Fundamentality

In this Section, we elaborate on the distinction between time reversal invariance which is a property of dynamical equations and time asymmetry which is a property of their solutions. If a solution of a dynamical equation is represented by a certain function of time $f(t)$, then we call it, in general, time-symmetric, if $f(t_0 + t) = f(t_0 - t)$ holds for some t_0. In particular, this is the case, if a solution describes *closed curves* in phase space. In particular, we can identify dynamical equations with physical laws and solutions to those equations with physical models that satisfy such laws.[15] In principle, there are four possible combinations of these properties:

(a) Time reversal invariance with only time-symmetric solutions.
(b) Time reversal invariance with some time-asymmetric solutions.
(c) No time reversal invariance with only time-symmetric solutions.
(d) No time reversal invariance with some time-asymmetric solutions.

---

[12]Price 1996, 88.
[13]Price 1996, 96–97.
[14]Price 1996, 97.
[15]Of course, not all dynamical equations describe physical laws, but physical laws can be understood as being described by dynamical equations.

Physical examples can be easily found for (a). On the other hand, since we are interested in time-asymmetric solutions, we shall not consider those examples. Examples for case (b) show that time-reversal-invariant laws (TRIL) can have time-asymmetric solutions; such can be found, for example, in Classical Electrodynamics. Whether case (c) and (d) are realized in physics, seems to be at least questionable; non-TRIL's cannot be found within the laws of fundamental physics in the standard interpretation; possible exceptions are some special versions of quantum laws, for example the rigged Hilbert space approach (see: Bohm, Gadella and Wickramasekara (1999); Bishop (2004), Castagnino, Gadella & Lombardi (2005), Castagnino, Gadella & Lombardi (2006)).

In the following, we shall show that case (b) represents a possible genuine way to define '*fundamentality*' for time-asymmetries, based only on the structure of the solution space of physical equations.

Definition I:
Suppose L is a fundamental time reversal invariant dynamic equation and S(L) is the solution space with dim(S(L)) = n. We will call S(L) *fundamentally time-asymmetric* if and only if

(i) There is some at most countable collection $S_i(L)$ of subspaces of dimensions $m_i < n$ and some at most uncountable collection $S_i'(L)$ of subspaces of dimension $m_i'+1<n$, such that: if $f(t) \in S(L)$ is time-symmetric, then $f(t) \in S_i(L)$ or $f(t) \in S_i'(L)$ for some i, and if $f(t) \in S(L)$ is time-asymmetric, then $f(t) \notin S_i(L)$ and $f(t) \notin S_i'(L)$ for all i.
(ii) For time-asymmetric solutions $f(t) \in S(L)$, the solution $f(-t) \in S(L)$ refers to the same physical world as f(t) does.

Condition (ii) is an important requirement. If $f(t) \in S(L)$ and $f(-t) \in S(L)$ would describe physically different worlds, then we would have to explain why only one direction (+ or − sign) occurs in nature. However, if condition (ii) holds, no such additional explanation is needed.

Moreover, in order to prevent a misunderstanding, we remark that the *combination* of both conditions is important. Condition (i) requires that 'almost all' solutions of a fundamental law are time-asymmetric. But this is not sufficient for an adequate notion of a fundamental time-asymmetry. The reason is that there can be dynamic equations such that for each time-asymmetric solution there exists a time mirrored twin solution which is time-asymmetric as well. Thus, condition (i) could be satisfied even if the totality of solutions would not show any preference of a particular time direction. Thus the notion of 'fundamentality' would be misplaced, since it could be applied even if the solution space would represent no directionality of time in nature.

Are there time reversal invariant physical laws with a solution space that is fundamentally time-asymmetric according to Definition I? The field equations of General Relativity indeed instantiate our definition, if we restrict the solution space in some physically well motivated way (see: Monecief (1975, 1976), Castagnino, Lara & Lombardi (2003a, 2003b)).

3. Applying the New Concept of Fundamentality to General Relativity

That the solution space of the field equations of Classical General Relativity, restricted in a physically well motivated way, is fundamentally time-asymmetric has been shown by Castagnino et al. (Castagnino (2003a), 374f, Castagnino *et al.* (2003b), 990f, see also Wohlfarth (2012)). The proof depends on the two following premises:

(i) The concept of cosmic time is not ruled out in the first place via the structure of the considered spacetime (as e.g. via being a non-orientable spacetime).[16]
(ii) The set of necessary dynamic variables contains further variables apart from the scaling factor.[17]

Even if spacetimes are mathematically possible in which the only dynamic variable is the scaling factor, such spacetimes are unphysical. The set of models (the spacetimes described by a particular solution of Einstein's equation) considered here should be the set of *physically plausible* spacetimes, i.e. the set of spacetimes which contain matter and energy as dynamical entities.

Given conditions (i) and (ii), we will show that time-asymmetry with respect to cosmic time is a typical property of spacetimes.

To start with: most of all *open* spacetimes are time-asymmetric. This follows from the fact that we can define the time-asymmetry of open spacetimes according to the asymmetric behaviour of the scaling factor as a function $a(t)$ of the cosmic time t: There exists, for open spacetimes, no hypersurface $t = t_S$ such that for all t: $a(t_S + t) = a(t_S - t)$. Hence, given such open time-orientable spacetime, it is obvious that spacetime geometry looks different in both temporal directions.[18] But, GR also allows *closed* spacetimes. Hence, in order to argue that even in the

[16] Since the condition of time-orientability guarantees that there is a consistent local time orientation for all points of spacetime, this weaker condition would be sufficient to reduce the class of spacetimes to those that have a unique local time order. But only the stronger condition of the existence of a cosmic time provides a global time function the value of which increases (decreases) along every timelike world line of the universe. Only then we can speak of 'two directions of time' for the whole universe.

[17] Notice, that Price's use of the Gold universe as a counterexample to any intrinsic time-asymmetry of the universe relies on his considering the scaling factor as the only parameter characterizing the universe.

[18] Note that there is one exception in classical cosmology: a static universe that also has an open topology but is symmetric (more precisely constant) with respect to the scaling factor. However, we will not consider the static solution of Einstein's equations because it requires fine tuning of the cosmological constant and the energy and matter content (and distribution) of the universe. Thus, according to classical cosmology, this solution is a special type that belongs to a subspace of solutions to Einstein's equation that has a lower dimension than the entire solution space.

set of closed time-orientable spacetime we find cosmic time-asymmetry, we have to analyse closed spacetimes in more detail (cf. Castagninio *et al.* (2003b), 900f.).

Our analysis will show that the set of time symmetric-spacetimes, even in closed topologies, is a set of measure zero. More precisely, this set of time-symmetric and closed solutions of Einstein's equation will turn out to have a lower dimension than the set of all closed solutions.[19]

For simplicity, consider an idealized case where the dynamics of spacetime is described by the scaling factor $a(t)$ and the scalar matter field $f(t)$ which depend on cosmic time t. In Hamiltonian mechanics, dynamic equations (the Hamiltonian) depend on dynamic variables and their first derivatives of the cosmic time parameter t. Thus, in this example, we have four variables in the Hamiltonian: $a(t)$, da/dt, $f(t)$, df/dt. Now, analytic mechanics always allows describing one of these variables as a function of the others. Thus, for simplicity, we have chosen $a(t) = f(da/dt, f(t), df/dt)$, where da/dt, $f(t)$, df/dt are now independent dynamic variables.

If we try to construct a symmetric spacetime, all dynamic variables must together behave in a time-symmetric manner. According to the singularity theorems in classical cosmology (Hawking & Ellis (1973)), we know that $a(t)$ has just one maximum. Next, we can choose the mathematical origin of cosmic time. For simplicity and without loss of generality, we stipulate the origin so as that $a(0)$ is the maximum value of the scaling factor. Thus, as a function of cosmic time, $a(t)$ is symmetric in relation to the axis a at the point $t = 0$, i.e. $a|_t = a|_{-t}$. Therefore, it is obvious that da/dt is symmetric in relation to the point $(t = 0; da/dt = 0)$, i.e. $\mathrm{da/dt}|_t = -\mathrm{da/dt}|_{-t}$.

However, for a time-symmetric spacetime, the behaviour of $f(t)$ and df/dt *together* with da/dt must also be symmetric. Thus, in this example, we have only two possibilities for the behaviour of $f(t)$ and df/dt at the cosmic time point $t = 0$, which makes the entire spacetime time-symmetric. These possibilities are given by the triplets $\{da/dt|_{t=0} = 0,\ f(t = 0),\ df/dt|_{t=0} = 0\}$, which is a symmetric solution of $f(t)$ about the f-axis at the point $t = 0$, and $\{da/dt|_{t=0} = 0,\ f(t = 0) = 0,\ df/dt|_{t=0}\}$ being a point-symmetric solution of $\varphi(t)$ with respect to the point $(t = 0;\ f(t = 0) = 0)$: $\varphi|_t = -\varphi|_{-t}$.

With respect to the general definition of time-symmetry of spacetimes (Section 1), the t = 0-axis represents the spatial hypersurface that splits the whole spacetime into "two halves that are temporal mirror images of each other". It does

[19]Spacetimes that have an open but time-symmetric (and not static) topology are open with respect to both past *and* future. We will not consider them because they require a change in the value of the cosmological constant. But, in the context of classical GR, the cosmological constant is *constant* in cosmic time. This may not be the case for full blown quantum or string cosmology, but these yet quite speculative accounts are beyond the scope of this study. In classical GR a contracting spacetime always includes a Big Crunch (see Hawking and Ellis 1973, Penrose 1979, Hawking and Penrose 1970). Thus, a spacetime cannot be open with respect to *two* directions of cosmic time if the spacetime is not static.

not matter, for that purpose, whether the respective function is symmetric about the a-axis (hypersurface) at the point t = 0 or whether it is point-symmetric with respect to the point t = 0. In both cases, the respective function develops in the same way for an observer starting at t = 0 and going in the direction of positive values of t as for an observer starting at t = 0 and going in the direction of negative values of t. The spacetime looks physically the same in both of these temporal directions, i.e. it is a symmetric spacetime.

All symmetric solutions can be constructed using the triplets given above. Thus, we can construct a subspace of time-symmetric solutions: *span* $\{(da/dt|_{t=0} = 0, f(t = 0), df/dt|_{t=0} = 0), (da/dt|_{t=0} = 0, f(t = 0) = 0, df/dt|_{t=0})\}$. The complete space for solutions of the dynamic equation, however, is given by *span* $\{(da/dt, 0, 0), (0, f, 0), (0, 0, df/dt)\}$

Thus, time-symmetric behaviour of a spacetime occurs only in a subset of solutions having a lower dimension than the entire set of solution *even* if we consider closed spacetimes.

The argument given above has shown that, assuming that cosmic time is definable and that we have more than the scaling factor to describe the dynamics of spacetime, time-asymmetry in terms of cosmic time is a typical property of the most idealized model. But this also holds true if we leave the very simple toy model of the example and add more dynamical variables, because the calculation in those cases would proceed in the same manner. The entire space of the solutions has always a higher dimension than that of the subset of time-symmetric solutions.

Despite the fact that the fundamental Einstein equation is time-reversal invariant, this does not imply that the models of General Relativity are time-symmetric. Instead, we have shown that *almost all* models, in which cosmic time can be defined, are time-asymmetric. Time-asymmetry is a typical property for the relevant set of solutions of the GR field equations.

It might be objected that every time-asymmetric model of GR has a time-reversed model which is also a solution of the Einstein equation and thus probably violates condition (ii). But this objection fails.[20] Two time-mirrored models, $f(t)$ and $f(-t)$, can be shown to represent the same physical world. We argue for that in the following three steps:

(a) Since the models are isomorphic, $f(t)$ does not possess intrinsic properties that are not possessed by $f(-t)$.
(b) The models describe spacetime as a whole. This implies that there is no time parameter (or any other physical parameter) outside the range of the geometrical objects $f(t)$ and $f(-t)$. The models are thus not related to a physical environment.

[20]We follow here the argumentation of Castagnino *et al.* (2009), 18.

(c) The combination of conditions (a) and (b) shows that two models $f(t)$ and $f(-t)$ do not differ in any intrinsic *or* extrinsic property. Thus, the models represent the same physical world.

It should be noticed that it is not only the case that *we* just cannot *distinguish* between the two solutions. The argument shows that there cannot be any property distinguishing between the two models.

4. The CPT-Objection

Earman has argued against Leibniz arguments of the form we used in Section 3. According to CPT invariance,[21] it follows from the assumption that $f(t)$ and $f(-t)$ describe the same physical situation that the combined change of C and P does not change the physical situation as well. But this cannot be true, Earrman argues, given that C and P are physically meaningful.

Again it will be important to take notice of the fact that spacetime has no background. The sign of a charge is clearly a conventional label. It is also clear that a positive and a negative charge in an everyday environment lead to different physical effects. In classical electrodynamics, the direction of the field lines is reversed if we change the sign of the associated electric charge. However, without a background, the direction of the field lines has no physical sense. We thus get a similar conclusion as with regard to the time-mirrored functions:

(i) The intrinsic properties of two electromagnetic fields that differ from each other only with respect to the sign of the electric charges, and thus also with respect to the direction of the field lines, are identical.
(ii) If there is no external environment, there are also no relational structures that differ with respect to C-mirrored functions.
(iii) Thus, both field configurations, which are formally distinguished by the signs of the charges, are physically identical, if there is no external environment.

Now essentially the same holds for the parity transformation P. The parity transformation that is used to formulate the CPT theorem (see for instance Gross (1999)) is a special type of Lorentz transformation given by $P(t) = t$ and $P(x) = -x$, where t is a time coordinate and x represents space coordinates. Thus, a parity transformation consists in the mirroring of space. Again, consider a solution of a global dynamic equation that describes the properties of spacetime as a whole, $f(t, c, p)$ (where t, c and p now refer to a time coordinate, the charge and the parity, respectively). The P transformation of this global solution is given by $P(f(t, c, p)) = f(t, c, -p)$. Here, the minus sign in front of p indicates that all space coordinates are mirrored.

[21] C is the charge transformation, P the parity transformation and T the time transformation for a particle. These transformations switch the signs of the charge C, parity P and time T. The CPT theorem which is empirically well confirmed says that physical situations do not change as the result of combined transformations of C, P and T.

However, they are mirrored according to the notion of some absolute (and thus not physical) coordinate system. In the context of general relativity, we clearly do not have absolute directions of space outside spacetime. Hence the above argumentation is repeated:

(i) All intrinsic properties of two spacetimes, $f(t, c, p)$ and $f(t, c, -p)$, are identical.
(ii) No external relationships exist for spacetimes, $f(t, c, p)$ and $f(t, c, -p)$.
(iii) Thus, the solutions $f(t, c, p)$ and $f(t, c, -p)$ are physically identical.

It has to be noticed that all this does not indicate anything about the possibility of one particular spacetime having the property of parity ('handedness') or the property of directedness of time. With respect to parity, this would mean that *within this* spacetime p and $-p$ differ from each other in a physical sense. However, the p-mirrored spacetime of such an intrinsically handed spacetime would have the *same* property of handedness and both spacetimes were thus identical. This shows that our argument does not exclude some intrinsic (versus relational) interpretation of parity.

In sum, we conclude that CP-transformations do not change any physical property of global solutions of the Einstein equations. Therefore, the CPT-objection is not successful against the result of Section 3: *Spacetimes of GR are fundamentally time asymmetric with respect to cosmic time and thus time directedness is a fundamental physical property in Classical Cosmology.*

Notice that the result depends on the application of the CPT-theorem to *spacetimes.* Within the context of particle physics, where the CPT-theorem has been established, we usually discuss the transformation of particle properties or systems constructed from particles, or, as in QFT, field-quanta. In such cases, a change of the P value is physically significant, as we observe in cases of parity violation in weak interactions.

5. Conclusion

Price's argument against the fundamentality of time-directedness has focused entirely on the lawlikeness interpretation of fundamentality. He has been right in arguing that, under this presupposition, no fundamental explanation of cosmological time-directedness can be provided. This remains true even when using a multiverse approach as theoretical background combined with some anthropic strategy; or when taking into account the new empirical findings about accelerated expansion of the universe.

Interestingly, Price himself had mentioned that symmetric theories may allow for asymmetric solutions. But he did not make use of that fact, assuming that the possibility of asymmetric solutions had to be based on some 'implicit' theoretical principles. We, instead propose an autonomous concept of fundamentality not based on fundamental dynamical equations, but on the solution space of such equations. This concept can be successfully applied to Classical Cosmology with the result that almost all solutions of the field equations (given adequate restrictions on the solution space) are time-asymmetric. This fact yields a fundamental basis for the time-directedness of our world.

References

[1] H. Price, *Time's Arrow and Archimedes' Point* (Oxford University Press, Oxford, 1996).

[2] H. Price, *Burbury's last case: The Mystery of the Entropic Arrow*, in *Time, Reality and Experience* (Cambridge University Press, Cambridge, 2002).

[3] H. Price, *The flow of time*, in *The Oxford Handbook of Time*, ed. Craig Callender, (Oxford University Press, Oxford, 2011), pp. 276—311.

[4] M. Castagnino and O. Lombardi, The global non-entropic arrow of time: From global geometrical asymmetry to local energy flow, *Synthese* **169** (2009), pp. 1–25.

[5] D.Z. Albert, *Time and Chance* (Harvard University Press, Cambridge, Massachusetts, 2000).

[6] M.M. Ćirković and V. Miloševic-Zdjelar, Three's A Crowd: On Causes, Entropy and Physical Eschatology, *Foundations of Science* **9** (2004), p. 1.

[7] A.D. Linde, *Inflation and Quantum Cosmology* (San Diego Academic Press, 1990).

[8] R. Penrose, Singularities and time asymmetry, in *General Relativity: An Einstein Centenary* (Cambridge University Press, Cambridge 1979).

[9] A.G. Riess *et al.*: Observational Evidence from Supernovae for an Accelerating Universe and a Cosmological Constant, *Astron. J.* **116**, 1009 (1998).

[10] J.G. Barlett and A. Blanchard, The Significance of the Cosmic Virial Theorem, *Astron. Astrophys.* **307**, 1 (1996).

[11] X. Fan, N.A. Bahcall and R. Cen, Determining the Amplitude of Mass Fluctuations in the Universe, *Astrophys. J.* **490**, L123, (1997).

[12] E. Bertschinger, Simulations of Structure Formation in the Universe, *Ann. Rev. Astron. Astrophys.* **36** (1998), p. 599.

[13] C.S. Kochanek, Gravitational Lensing Limits on Cold Dark Matter and Its Variants, *Astrophys. J.* **453**, (1995), p. 545.

[14] P. Coles and G. Ellis, The Case of an Open Universe, *Nature*, **370**, (1994), p. 609.

[15] S. Goldstein, Typicality and Notions of Probability in Physics, Manuscript (2011).

[16] T. Maudlin, What could be objective about probabilities?, *Studies in History and Philosophy of Modern Physics* **38** (2007), pp. 275–291.

[17] L. Boltzmann, Über die Beziehung zwischen dem zweiten Hauptsatze der mechanischen Wärmetheorie und der Wahrscheinlichkeitsrechnung respective den Sätzen über das Wärmegleichgewicht, *Sitzungsberichte der Akademie der Wissenschaften zu Wien* **76** (1877), pp. 373–435.

[18] A. Bohm, M. Gadella and S. Wickramasekara, Some little things about rigged Hilbert spaces and quantum mechanics and all that, in *Generalized Functions, Operator Theory, and Dynamical Systems*, eds. I. Antoniou and G. Lumer (Chapman & Hall/CRC, Boca Raton, 1999), pp. 202–250.

[19] R.C. Bishop, Arrow of Time in Rigged Hilbert Space Quantum Mechanics, *Int. J. Theor. Phys.* **43**, 7/8 (2004).

[20] M. Castagnino, M. Gadella, and O. Lombardi, Time's arrow and irreversibility in time-asymmetric quantum mechanics; *International Studies in the Philosophy of Science*, **19** (2005), pp. 223–243.

[21] M. Castagnino, M. Gadella, and O. Lombardi, Time-reversal, irreversibility and arrow of time in quantum mechanics; *Foundations of Physics* **36** (2006), pp. 407–426.

[22] V. Moncief, Spacetime symmetries and linearization stability of the Einstein equations. I, *Journ. Math. Phys.* **16** (1975), p. 493.
[23] V. Moncief, Spacetime symmetries and linearization stability of the Einstein equations. II, *Journ. Math. Phys.* **17** (1975), p. 1893.
[24] M. Castagnino, L. Lara and O. Lombardi, The cosmological origin of time asymmetries, *Classical and Quantum Gravity* **20** (2003a), pp. 369–391.
[25] M. Castagnino, L. Lara and O. Lombardi, The Direction of Time: From the Global Arrow to the Local Arrow, *International Journal of Theoretical Physics* **42** (10) (2003b), pp. 2487–2504.
[26] D. Wohlfarth, A New View of "Fundamentality" for Time Asymmetries in Modern Physics, Proceedings of the EPSA 11 conference in Athens, October 2011: Recent Progress in Philosophy of Science: Perspectives and Foundational Problems (Springer, New York, 2012).
[27] S.W. Hawking and G.F.R Ellis, *The Large Scale Structure of Space-time* (Cambridge University Press, Cambridge 1973).
[28] S.W. Hawking, R. Penrose: Proc. R. Soc. A. 314 (1970).
[29] J. Earman, An Attempt to add a little Direction to "the Problem of the Direction of Time", *Philosophy of Science* 41 (1974), pp. 15–47.
[30] F. Gross, *Relativistic Quantum Mechanics and Field Theory* (Wiley, New York 1999).

EXPLANATION, SPECIAL RELATIVITY, AND PRESENTISM

KATHERINE BRADING

Department of Philosophy, University of Notre Dame,
Notre Dame, IN 46556, U.S.A.
kbrading@nd.edu

Which aspects of our experience of time can special relativity be used to explain? In this paper I focus on those aspects of our experience of time associated with presentism. Special relativity has been used to argue against presentism, and to insist that those aspects of our experience of time that are central to the presentist thesis are to be "explained away" as somehow illusory. In this paper I offer a version of special relativistic presentism that has the potential to explain (rather than explain away) these aspects of our experience of time, including the short duration of the present and temporal becoming.

Keywords: Special relativity; presentism; law-constitutive.

1. Introduction[1]

Presentism, as I use the term in this paper, is the thesis that the temporal extent of what exists is less than the age of the universe. This is a weak thesis, and many presentists would wish to strengthen it in various ways. But I take this thesis to be something that no presentists would deny, while it would be rejected by their opponents, the four-dimensional eternalists, and for this reason it will prove to be a useful tool in what follows.

In fact, presentists typically take the present to be *very* short compared to the age of the universe. For example, some presentists might suggest that the present has *no* temporal extent at all, being merely the boundary between the temporally extended past and the temporally extended future. Others may admit some temporal extension for the present, perhaps using the idea of the "specious present" (the idea that our experiences, to be experiences at all, must have a duration) as motivation (our experience of the present is of a temporally extended present, and this is what we should be accounting for in our metaphysics), but in all cases the present is considered to be very short in comparison to the age of the universe. The spatial extent of the present, by contrast, is typically considered to be as big as space itself.

[1]I am grateful to the organizers and hosts of SEMS 2012 for their warm hospitality. My talk was, and therefore this paper is, related to work published in *Studies in History and Philosophy of Science* and *Philosophy of Science* (see Brading, 2013b and forthcoming).

It is the ontological status of this temporally small and spatially large present that is at stake in much of the dispute between presentists and four-dimensionalists.

Special relativity has been used to argue against presentism and on behalf of four-dimensional eternalism, and we will look at the central argument below (see section 3). In this paper, I will argue against the mainstream: I will argue for the possibility of a special relativistic version of presentism.[2]

Special relativity demands radical conceptual revisions; there is no escaping this. Nevertheless, I maintain that the core of the presentist thesis is preserved in the view that I offer.[3] It may be that the view can be empirically refuted, or that conceptual revisions of the overall package make the price of presentism too high, but I set these issues to one side for the purposes of this paper. My goal in this paper is to show one way in which special relativity might be used to provide an explanation of aspects of our experience typically associated with presentism, and typically rejected as illusory by four-dimensional eternalists.

2. Explanation and Special Relativity

There are various aspects of our "experience of time" that special relativity, as standardly understood by philosophers of physics, can be used to explain. We take the Minkowski structure of spacetime, and we look to see which things have correlates in that structure and which things don't. Those things that *do* have correlates are candidates for being explained by special relativity: we have identified (via our physical theory) an ontological correlate for an element of our experience of time and we have thereby met a necessary condition for explaining that element of our experience using that particular theory. For those aspects of our "experience of time" that *do not* have correlates in special relativity, we must decide whether the theory *explains them away*, or is just *silent about them*. Here is an example of each of the three possibilities:

- Temporal intervals between events are a feature of our experience of time (via our use of clocks, for example, or simply through our own biological rhythms) that is *explained*. The structure of Minkowski spacetime provides an absolute spacetime interval between any pair of events, and the spacetime interval as measured along the worldline of an inertially moving object is a temporal interval. That is to say, an inertially moving clock measures the temporal interval along its worldline. Your experience of temporal intervals can be explained by appeal to the temporal intervals along your worldline, themselves part of the structure of Minkowski spacetime. In this sense, our experience that there are temporal intervals can be explained by appeal to special relativity.

[2]This paper should be read in the context of the literature that includes Hinchcliff (2000), Savitt (2000), Gibson and Pooley (2006) and the papers referred to therein.
[3]I will say what I take this core to be and why as we move on.

- That time has a direction is also a feature of our experience of time, but special relativity is itself *silent* about this directionality. The structure of Minkowski spacetime is such that we *can* add a directionality to timelike curves if we want to (we can label one direction "future" and one direction "past"), but nothing in special relativity itself requires us to do so. Rather, a direction of time is usually built into Minkowski spacetime by hand. For this reason, it is not clear that there is any explanatory work being done by the structure of Minkowski spacetime when it comes to a direction of time. We put it in because our experience, and indeed other areas of physics, seem to indicate we need it, but special relativity itself is silent about this feature of our experience of time.
- When it comes to a flow of time there is no correlate in the structure of Minkowski spacetime. While the idea of a "flow" of time is notoriously hard to make precise without falling into contradiction or regress, the standard philosophy of physics conclusion is straightforward. Since there is nothing in Minkowski spacetime that could serve as a correlate for any such "flow", the flow of time is to be *explained away* as an illusion that has no counterpart in the spatiotemporal structure of the world.[4]

My view is that our explorations of our spatiotemporal concepts through developments in physics *can* be used to help us distinguish between aspects of our "experience of time" which have an explanatory basis in the spatiotemporal structure of the world versus those which are to be "explained away". However, I wish to challenge the standard way in which special relativity is used to make these judgements, and what the outcome of this process is usually taken to be. As noted above, the focus of my discussion will be so-called "presentism", and the claim that "the present" is an illusion to be explained away.

3. Special Relativity Against Presentism

In this section we review the standard argument against presentism (section 3.1), and the possibilities for explaining the present by means of Minkowski spacetime (section 3.2). While the case looks rather hopeless, we will take from this discussion one of the clues that we need for developing a special relativistic presentism in the remainder of the paper.

[4]One temptation for advocates of "flow" might be the "spotlight" metaphor according to which a spotlight moves along a worldline, progressively lighting up successive sections of the worldline, and this represents the "flow" of time. This demonstrates the tendency towards regress, since the addition of a moving spotlight can be represented through the addition of a fifth dimension (a second temporal dimension) leading to a five-dimensional block in which there is no flow, along with the "appearance" of flow in the reduced four-dimensional space. Additional puzzles concern the relationship between the original temporal dimension (representing temporal intervals etc.) and the newly-introduced fifth dimension (representing "flow" of time).

3.1. *The standard argument*

Let us begin with the apparent clash between "our experience of time" and special relativity. I will characterize what the standard options are, and then (in section 4, below) say why I think there is another option on the table. Above, I said that presentism is the view that the temporal extent of what exists is less than the age of the universe, and that typically a stronger view is involved. The stronger view asserts that all and only the things that exist *now* are real where "now" is taken to be of *very* small temporal extent compared to the age of the universe, perhaps being of infinitesimally short duration, if not instantaneous.

By itself, this is insufficient to generate a version of presentism that should be "explained away" in the light of special relativity, but it does give rise to a very peculiar version of presentism, as Howard Stein (1968) notes in one of the papers that is now part of the canonical literature on special relativity and presentism.

The argument runs as follows. Take an event in spacetime, say e1. assume that e1 is "now" relative to itself. The other events in spacetime that are "now" relative to e1 are those which are *simultaneous* with e1. But, as is well known, simultaneity of distant events is precisely the notion that is problematized by Einstein in his 1905 paper in which he sets out his special theory of relativity. Early in the "kinematical part" of the paper, Einstein argues that in order to determine simultaneity of distant events we must make use of a synchrony convention. For some philosophers of physics, conventionality of simultaneity implies that *any* "joining of the dots" in planes of simultaneity involves a failure to appreciate the philosophical lessons of special relativity: there is no (non-conventional) fact about which events are now with respect to e1 (other than e1 itself), therefore no events other than e1 are determinately now relative to e1. For other philosophers of physics, it is not the conventionality that yields this result, but the relativity of simultaneity that follows from adopting the Einstein synchrony convention in setting up all inertial coordinate systems. The synchrony convention is held not to be problematic in itself, on this version of the argument, but the consequent relativity of simultaneity has the upshot that which events are simultaneous with e1 varies from one inertial coordinate system to another (moving with respect to the first). Once again, there is no unique set of events larger than e1 itself where those events are determinately now with respect to e1.

One conclusion we might draw is that only e1 is "now" relative to itself. This would not be incompatible with presentism as I have characterized it above. The temporal extent of the present relative to e1 would indeed be very short, since it would be restricted to e1 itself and e1 is a spatiotemporal point. Such a view has some other features worth remarking: the present is not *the* present, it is "the present relative to e1", and it is very small *spatially* as well as temporally (it is restricted to a spacetime point). Moreover, if we add the claim that what is real is what is now, our presentism leads us, as Stein (1968) says, to "a peculiarly extreme (but pluralistic) form of solipsism." What *exists*, relative to e1, is just e1 and nothing else. This is not a very attractive form of presentism, as I think we can agree.[5]

There are two other standard options available, given the argument from special relativity.[6] There are those who suggest that taking our experience of time seriously requires us to reject, or supplement, special relativity. We might try to stay as close as possible to special relativity, but add a preferred foliation so that we get a unique, global "now". On this view, the present is very short temporally (it is a surface) and very large spatially (*every* timelike or lightlike curve in our spacetime cuts "the present" exactly once). So now let's return to our three possibilities for the explanatory power of special relativity.

3.2. *Explaining the present in Minkowski spacetime*

If we adopt the approach of adding a preferred foliation, does special relativity explain the present, is it silent about the present, or does it explain away the present?

Consider first the possibility that special relativity can be used to *explain* the present. In the account offered above, special relativity explains temporal intervals via the spatiotemporal intervals that are part of the structure of Minkowski spacetime. However, there is nothing analogous in the structure of Minkowski spacetime by which special relativity can be used to *explain* the present construed as a preferred foliation: Minkowski spacetime does not come foliated, let alone with a preferred foliation. On the other hand, there are those who suggest that we should give up presentism and adopt the block universe. So special relativity cannot *explain* the present in a manner analogous to the explanation of temporal intervals.

Advocates of presentism do not want to use special relativity to "explain away" our experience of "the present", so the hope must be that special relativity is simply silent about the present, and that we are free to add the "present" just as we are free to add a temporal direction (see above). There is, however, an important disanalogy between the case of temporal direction and the case of the present, relevant to these purposes: choosing a temporal direction for Minkowski spacetime is tied to our experiences in way that choosing a preferred foliation is not, as we shall see. Recall first the case of temporal direction. While there is nothing in the structure of Minkowski spacetime that corresponds to a temporal direction, we have independent empirical access to such a direction so that we are able to align our representations of events in Minkowski spacetime with our experienced direction of time (our representation agrees with our experience of before-after relations among events along a given worldline, for example). Having chosen a "forwards" direction

[5]See Hinchcliff's (2000) pp. S579–80 discussion of "point presentism", in which he reviews an additional objection made by Putnam (1967) and responded to by Stein (1968).

[6]For a more detailed discussion, additional options, and references to the literature, see Savitt (2000), which provides an excellent entry into the large literature on this topic, while developing Savitt's own position.

along a timelike curve in Minkowski spacetime, we have thereby temporally oriented our representation of events in the whole of Minkowski spacetime, and our experience of temporal direction will not only be consistent with this global temporal orientation, but also *inconsistent* with making different choices of orientation at different times and places in our spacetime. In other words, we can add a direction of time to Minkowksi spacetime in an empirically significant manner; the direction we choose is tied to our experiences of the direction of time.

Now compare this to the case of a preferred foliation. As with the case of temporal direction, there is nothing in the structure of Minkowski spacetime that corresponds to a preferred foliation, but also nothing to stop us adding one. So far, so good. However, unlike in the case of temporal direction, we do not have any independent access to such a preferred foliation, and so we have no means of aligning our attempted representation of "the present" by means of a preferred foliation with our experience of "the present". Choosing a preferred foliation has no tie to experience: there is *nothing* in our experience that enables us to determine *which* of the infinitely many such surfaces passing through a given point is "the" present. Moreover, unlike in the case of temporal direction, we have complete freedom to choose different foliations at different times and places without any consequences for our experiences whatsoever. As Savitt (2000, p. S570) writes, "If the present is indeed so elusive, I find it difficult to imagine what aid or comfort it could be...". For our purposes, we are trying to do justice to our "experience of time", and since a "present" constructed by means of imposing a preferred foliation leaves *no trace of any kind* in our experience, it is of no use to us. Special relativity is not merely silent about our experience of the present: there is no means by which to add this experience to the structure of Minkowski spacetime by using a preferred foliation. Our attempt to represent our experience of "the present" in Minkowksi spacetime by adding a preferred foliation fails.

This brings us to the second of the standard options, which is to abandon presentism in the light of special relativity and endorse the four-dimensional "block universe". This is the dominant view amongst philosophers of physics. According to this view, there is nothing in special relativity, or in the structure of Minkowski spacetime, that corresponds to "the present". Moreover, advocates of this view accept the conclusion of the preceding paragraph: the structure of Minkowski spacetime is such that we cannot "add" a present. Special relativity is not silent about "the present"; on the contrary, the lesson of special relativity is that "the present" is an illusion that should be explained away.[7]

The underlying reason is, of course, the transition from space and time to spacetime. As Gibson and Pooley (2006, p. 160) stress, in reference to Savitt (2000), "The reason why there is no time like the present in Minkowski spacetime is that,

[7]For one approach to this, see Barbour (1999). This approach not only explains away the present, but also directionality and even temporal order!

in Minkowski spacetime, *there is no time*." What we have, instead, is spacetime, and regions of spacetime. We might take this to be decisive against presentism, but a different lesson is also possible: whatever the present might be, it cannot be a time, it must be a spatiotemporal region.

4. What is Present and What is Real

Our conclusion so far is this: whatever the present might be, it cannot be a time, it must be a spatiotemporal region. In order to construct such a present, our first step will be to re-examine the existence claims associated with presentism and with the block universe. We will then be in a position to develop a special relativistic version of presentism.

Presentists maintain that the temporal extent of what exists is restricted to what is present. Thus, the present plays a crucial ontological role: the spatial extent of what is real depends on the spatial extent of what is present. For example, if there is absolute simultaneity, then the spatial extent of the present is as large as the universe itself, and what is real relative to me-now extends spatially throughout the universe. But, as is familiar and as we discussed above, special relativity does not support absolute simultaneity, and so any version of presentism built on absolute simultaneity is doomed from the perspective of special relativity. The alternative presentist position that we were offered is extreme pluralistic solipsism in which the present relative to a given event, and thereby what is real relative to that event, is restricted to that event itself. Or, to put it more personally, what is real for "me-now" is nothing over and above me-now: *nothing else is real for me*. We shall seek a special relativistic presentism that does not lead to solipsism.

The only hope for such a presentist, I suggest, is to break the connection between what is present and what is real. This can be done by relegating spacetime to an epistemic rather than an ontological status and appealing instead to the dynamical laws for accounts of what is present and what is real, as we shall see.

We can understand the epistemic role of spacetime by paying attention to the following considerations. The reason that we set out an arena of space and time when we're doing physics is that part of what we're trying to do is to extend our knowledge of events to times and places distant from the here and now. Given initial conditions for a particle, for example, we want to know its future trajectory: we want to know what other places it will go to and at what other times it will be there. We construct a theoretical arena of space and time and plot the trajectory of our particle in that arena, using this as a means of making predictions about the location of the particle at future times: we are attempting to extend our epistemic reach beyond the here-now. Moreover, we try to unite our predictions for different physical systems within a single spatiotemporal arena, as part of our efforts to organize our knowledge of the not-here and/or not-now. For certain kinds of systems, physics has been very successful in this project, and it can be tempting to ontologize the spatiotemporal arena. For our purposes, however, the important point is this: there is no necessary inference from the above epistemic role for spacetime to an ontological status for the spatiotemporal arena. In the case of special relativity, the success of special relativistic physics does not in itself warrant the inference to a block universe ontology.

In sections 2 and 3 above we appealed to the structure of Minkowski spacetime in asking what features of our experience of time can be explained (or should be explained away) using special relativity. In the following sections, we take a different approach in which we appeal to dynamical laws, rather than to Minkowski spacetime, as our guide to ontological commitments. At the heart of special relativity are the Lorentz transformations: Minkowski spacetime is a geometrical structure encoding the invariant quantities associated with the Lorentz transformations of special relativity, and special relativistic dynamical laws have the Lorentz transformations as symmetries of those laws. There is disagreement in the literature over how to prioritize the laws versus the spacetime structure (see Van Camp, 2011, for example, and references therein to a debate whose main protagonists are Harvey Brown and Michel Janssen). We will not enter into this debate here. Rather, having discussed how presentism fares in the light of Minkowski spacetime, we will now examine presentism under the requirement that our laws be Lorentz invariant.

5. A Law-Constitutive Approach to What is Real and What is Present

In this section, we start from the dynamical laws, rather than the structure of space and time, in order to develop accounts of what is real and what is present. This allows us to formulate a special relativistic version of presentism.

5.1. *The dynamical laws and what is real*

I advocate what I call a law-constitutive approach to the ontology of physics. I have argued for this in the context of Newton's *Principia*, but the approach generalizes.[8] In the case of Newton, my claim is that his laws of motion and force laws play a role in constituting the very things that are their subject-matter. Since the laws explicitly concern "bodies", the claim is that we don't first have bodies and then ask what laws they satisfy; rather, in specifying the laws we thereby characterize the "bodies" to which they apply. For example, Newton's first law says that "Every body perseveres in its state of being at rest or of moving uniformly straight forward, except insofar as it is compelled to change its state by forces impressed", and from a law-constitutive perspective what this means is that *part of what it is* to be an isolated thing *is* for it to conserve its quantity and direction of motion. Newton's second and third laws extend his first law to composite systems, ensuring that such

[8]See Brading (2012, 2011, and 2013a). See also Brading and Skiles (2012).

composites themselves obey the first law. From a law-constitutive perspective, what we learn is that to be a body, be it simple or composite, *is* (in part) to satisfy the law of conservation of motion. In this way, Newton's laws play a crucial role in constituting what exists. More generally, a necessary (and sufficient, in the strong version) condition for being a physical thing is satisfying the laws of a given physical theory. This is not the weak (and perhaps uncontroversial) claim that to be a certain *kind* of thing is to satisfy a certain system of laws. Rather, the claim is that what it is to be a physical thing *at all* is to satisfy a certain system of physical laws: the laws are constitutive of what exists.

An important point for our purposes is this: where satisfaction of a set of dynamical laws requires that a temporally extended process take place, according to the law-constitutive approach the very system that is the subject-matter of those laws *does not exist* on timescales shorter than the minimum length of that process. We will come back to this point below.

The restriction imposed by special relativity is that the laws must be Lorentz invariant. Suppose, then, that I have a Lorentz invariant system of laws, and that I have an object (let's be generous, and personalize this, by making it a person, me) that satisfies that system of laws. I am, on the law-constitutive approach, therefore real; I exist. What else is real for me? Physics offers us at least two ways to go: we can use spacetime structure to answer this question (e.g., what is real for me is what is present for me), or we can use the dynamical laws. Since we have used the laws to give an account of my own existence, let's follow that approach further and use it to address our new question.

In the Newtonian case above, we went from a single body (addressed in Newton's first law) to a composite system of bodies *via interactions* (addressed in his second and third laws). A natural approach to adopt, therefore, is an appeal to law-governed interactions: what is real relative to me are just those things with which I am interacting.

In the Newtonian cosmos of instantaneous gravitational interaction between all masses, the two approaches (using spacetime structure and using interactions according to dynamical laws) coincide. For all massive bodies to interact *instantaneously* with one another, the spacetime structure must include absolute *simultaneity*. Thus, "real for me" defined in terms of absolute simultaneity coincides with "real for me" defined in terms of interactions: at a given moment[9] in my life, all bodies existing at that absolute moment are present (and therefore real relative to me, according to the spacetime approach), and are interacting with me (and therefore real relative to me, according to the dynamical laws approach). In special relativity, however, we can (and should) tease these things apart.

[9]Here, we slide over the important point made above: if we follow the law-constitutive approach it is likely to turn out that I don't exist at all at an instant, requiring instead a temporally extended "moment". We return to this point below.

5.2. *The dynamical laws and what is present*

The approach we are taking is to try treating Minkowksi spacetime as playing a primarily epistemic role: while matter is understood as being spatiotemporal, it is not thought of as being *in* space and time in the sense suggested by the block universe.[10] Instead, whatever ontological commitments we have concerning space and time come from paying attention to the details of the dynamical laws of matter, rather than to Minkowski spacetime, and according to special relativity these laws must be Lorentz invariant. Thus, as a first step, we are seeking to determine whether we can obtain an account of "the present" from the requirement of Lorentz invariant laws. As noted above (see end of section 3), one lesson from special relativity is that, if there is to be a spatially extended present, it must be a spatiotemporally extended present. The same reasoning (that in special relativity space and time are unified into spacetime) applies to spatiotemporally extended things: for any things that are spatiotemporally extended, within the spacetime region of those things there are no purely spatial or purely temporal relations. Rather, things "occupy" time just as they "occupy" space: by being spatiotemporally extended.

If we now appeal to the dynamical laws we can say that the present of a given system is a spatiotemporal region of whatever size is required according to those dynamical laws in order to constitute the system in question.[11] For example, as noted above, where satisfaction of a set of dynamical laws requires a temporally extended process, according to the law-constitutive approach the very system that is the subject-matter of those laws *does not exist* on timescales shorter than the minimum length of that process. Thus, on this approach, the size and shape of "the present" for a given dynamical system becomes an empirical question dependent on the details of the dynamical laws for that system. (Moreover, I can be made of things each of whose dynamically sustained region of spacetime (i.e. the "present" of each) is much smaller than mine, so that my constituents come into and go out of existence on much shorter timescales than I do. But let's set aside that issue for the purposes of this paper.)

It might turn out that the extent of the spatiotemporal region necessary to sustain any given system needs to be the *entire* spatiotemporal history of the universe. Should that indeed turn out to be the case, we will find ourselves with a "present" that is the entire block universe, and this is no longer presentism as I defined it at the beginning of this paper. But things might not turn out that way. If the size of the spatiotemporal region required to sustain a given dynamical system is less than

[10]Thus, space and time are not independent of matter ontologically (this is not any version of substantivalism), but neither do the spatiotemporal characteristics of matter reduce to relations among material things (so this is not a version of relationism either); rather, the relationship between space, time and matter is a middle way between absolute space and time versus relationism.
[11]There are no determinate *ontological* spatial or temporal relations within that region, and that system stands in no determinate *ontological* spatial or temporal relations to any other system. All the ontology is carried by the dynamics; we frame the dynamics spatiotemporally, but ontologically the dynamics requires no such spatiotemporal underpinning.

the entire spatiotemporal evolution of the universe as represented in Minkowski spacetime, then "now" for that system is smaller than the block universe, in both its spatial and its temporal extent, and this satisfies the definition of presentism that I gave at the start of the paper. Notice that any such present will be spatially smaller than the size of the universe, as well as temporally shorter; this is one of the lessons of special relativity and of the unification of space and time into spacetime.

That the size of the spatiotemporal region required to sustain a dynamical system is less than the entire spatiotemporal evolution of the universe as represented in Minkowski spacetime satisfies the definition of presentism given at the start of this paper, but one reason that it might seem too weak is that "the present" associated with a given system might encompass the entire lifetime of that system. If "my present" turns out to include my entire lifetime, then this version of presentism does not offer a "present" that correlates with our "experience" of the present, and this is what we were seeking as part of our explanatory strategy.

Let us return to the notion of the "specious present", which is the idea that our experiences, to be experiences at all, must have a duration. Following Gibson and Pooley (2006, p. 166), let's say that this duration is about 0.2 of a second along your worldline. Taking into account your spatial extent, there is a spatiotemporal region that you occupy corresponding to the "specious present" which Gibson and Pooley call *NOW*. This *NOW* has been arrived at from our experience of time, and for Gibson and Pooley it is situated within an eternalist view of spacetime. The proposal I am offering here begins instead from the dynamical laws relative to which you constitute a physical system; it seeks the physical underpinnings of the "specious present", and seeks to define your present as the spatiotemporal region of whatever size is necessary for your existence as a physical system. Insofar as the "specious present" is a manifestation of the size of the spatiotemporal region needed to sustain your existence, then we can expect the size of this region to be at least as big as *NOW*. If we are able to connect the "present" as derived from Lorentz invariant dynamical laws to the "specious present" of our experience, then special relativistic physics will contain the resources for *explaining* the present.

Before turning to a formulation of special relativistic presentism along these lines, I first need to return to the issue of solipsism, which I said our version of presentism would avoid.

5.3. *Escaping solipsism*

Typically, presentism is formulated in terms of what is real, as for example in the statement "All and only things that exist now are real". According to the proposal being developed here, "the present" is defined as the size of the spatiotemporal region necessary to sustain the dynamical system in question. Thus, "the present" is defined relative to a given system, and for that system alone. If "the present" is relative to a single system, then for each such system there is a present, but if there is no shared present then we are back to a form of pluralistic solipsism.

What we are learning from our considerations of special relativity is that the connections between what is present, what is real, and what exists, are more complex than the above statement of presentism allows. According to the account being offered here, "the present" concerns only the existence of a given system, and plays no role in "extending" existence from that system to other systems. In order to avoid solipsism, the account must be supplemented with something else that supports existence claims across multiple systems, and in fact it does so naturally, through interactions. What is real (i.e. what exists), relative to me, are all those systems with which I am interacting. Thus, the existence of a given system is characterized in terms of its present, and the existence of other systems relative to that system is characterized in terms of interactions and *not* in terms of a shared present. To see the force of this distinction, consider the following example. Do my sons exist relative to me, when they are in Indiana, U.S.A., and I am at Shanxi University in China? Of course. There are plenty of interactions going on that link us. But is there a determinate fact of the matter about what they are doing *right now*? No. What makes the rest of the world real relative to me-now is not that it exists "now" (this claim has been rendered meaningless by special relativity), but that I am interacting with it. Being "real" is a relation that holds between dynamical systems: we belong to the same world as whatever we interact with, and the rest of that world is as real (relative to us) as we are.

Of things not spatiotemporally coincident with me, I can ask "What else is real, relative to me-now?", but I cannot ask "What else is now, relative to me-now?" because only *I* am now relative to me-now.

Interesting questions arise over the spatiotemporal extent of what is real relative to me-now, including whether this relation of being real is symmetric: do we have to have *inter*action (i.e. two way) for reality, or just action of one thing upon another? If only action, then the relation of being real fails to be symmetric, and furthermore we might wonder how diluted the causal chain has to get before reality fails. That we can raise these kinds of questions might seem attractive: there's a sense in which the supernova of 1572 exists for me, and a sense in which it doesn't (on the one hand the light from it reaches me, so the supernova acts on me via my interaction with that light, but on the other hand I don't *interact* with the supernova (nothing emanating from me reaches the event of the star exploding)), and furthermore there seems to be no sense in which I exist for the supernova of 1572. The spatiotemporal structures relevant to these distinctions can be represented using Minkowski spacetime. For example, if the region of my present corresponds is Gibson and Pooley's *NOW*, then those things with which I am *inter*acting can be represented as falling within what Gibson and Pooley (2006, p. 167) refer to as the *Stein Present* of *NOW*: the spatiotemporal region bounded by the forward lightcone of the past boundary of *NOW* and the backward lightcone of the future boundary of *NOW*. Gibson and Pooley point out that this region is very thin temporally, but large spatially, with the "wedge" around the *NOW* extending out to a diameter of about 60,000 km. A region of at least this size is interactively real to me, and as a matter of our experience of time this region might, as they point out, "be easily mistaken for an instantaneous, global present" (Gibson and Pooley, 2006, p. 167).[12]

We have developed the outline of a special relativistic version of presentism, based on the requirement that our dynamical laws be Lorentz invariant. In the following section we flesh out the position a little more.

6. Special Relativistic Presentism

We have outlined a special relativistic presentism that is based the requirement that our dynamical laws be Lorentz invariant.The weakest version of this view, compatible with presentism as the thesis that the temporal extent of what exists is less than age of the universe, says:

> For each and every thing, that thing exists only presently, where the spatiotemporal extent of the present associated with a given thing is the spatiotemporal region necessary to sustain that system as determined by the dynamical laws governing that thing, and where that spatiotemporal region is less than the entire spatiotemporal evolution of the universe as represented in Minkowski spacetime.

As we have seen, we have to complexify our account of "what exists" to distinguish between my existence (which is the focus of this characterization of presentism) and the existence of other things relative to me (which we do not account for using presentism, or any aspect of spatiotemporal structure, but via interactions). This enables us to avoid solipsism. However, the above characterization is too weak. In order to recover our experience of "the present", we need to pay attention to people and to strengthen this last clause appropriately. Call this strengthened version Special Relativistic Presentism (SRP):

> **(SRP)** For each and every thing, that thing exists only presently, where the spatiotemporal extent of the present associated with a given thing is the spatiotemporal region necessary to sustain that system as determined by the dynamical laws governing that thing, and where that spatiotemporal region for people is approximately the size of the specious present as represented in Minkowski spacetime.

Such a present is temporally thin and allows for temporal becoming. Moreover, with interactions taken into account, "what is real" relative to me-now is similarly temporally thin but also spatially extended to include everyone else on this planet. We can supplement SRP with the following condition on extending what exists

[12] Whether or not relative reality is symmetric is an open question. What is not open is whether relative reality is transitive: it is not (just because A interacts with B, and B interacts with C, it does not follow that A interacts with C).

beyond me-now to include other things, call this special relativistic reality (SRR):

> **(SRR)** For any given present thing, all those things with which it is interacting are real (exist) with respect to that thing.

I take these features, of a temporally thin me-now that admits of temporal becoming, and a spatiotemporally extended reality relative to me-now, to capture the core of the presentist thesis as it relates to our "experience of time".[13]

If SRP succeeds, and we adopt SRR, then special relativistic physics does not require that we explain away those features of our experience of time associated with presentism, and nor is it silent about them. On the contrary, it explains them.

Some final clarifications are in order. Notice that SRP is an *empirical* hypothesis. Should the hypothesis be borne out, we will be able to use special relativity to explain our experience of the present as temporally thin and spatially extended, along with becoming. Thus, the possibility remains on the table that the core of the presentist thesis can be legitimately conserved while taking the lessons of special relativity fully on board via Lorentz invariance of the laws. The presentist must revise her position in the light of special relativity, but she doesn't have to give up the core. Notice, too, that in order to explain the above "presentist" aspects of our experience of time, more than special relativity is required. We need to use the *details* of Lorentz invariant dynamical laws in order show that the condition SRP is met. The spatiotemporal extent of the present is an empirical matter, on this approach, which lives or dies according to the details of the laws of physics as they are developed. For example, attention to the dynamical laws in the manner required by SRP may lead us back to the block universe. But the following point stands: this ontological conclusion is something that cannot be reached directly from the Lorentz invariance of those laws. Another way in which SRP might fail empirically is if there turn out to be no such things as people on our law-constitutive approach to ontology. But again, this is an empirical question, to be settled not by special relativity alone but by the details of Lorentz invariant dynamical laws.

In sum, we have developed a version of special relativistic presentism that does no violence to special relativity itself (we do not add a preferred foliation of spacetime, or anything like that), that does not lead to solipsism, and in which special relativistic physics explains those aspects of our experience of time which lie at the heart of the presentist position.

---

[13]Hinchcliff (2000, p. S580) advises against trying to identify a "core" of presentism. I agree that we should not look to the literature for any such "core", but I do think we can retreat to what it is in our experience that we are seeking to do justice to in advocating presentism, and ask whether we can construct a position that explains those features of experience rather than explaining them away (as illusions, even if well-founded illusions).

7. Final Remarks

That we can systematize things in a global spatiotemporal framework is surely an interesting fact, and you might want to ontologize the overall framework above and beyond the dynamics. You can if you want to. My point has been that you don't have to: an alternative approach is available.

Why care? Well, for one thing I think it's fun to explore the space of options, even the wacky ones like this. But let me offer two further motivations. One is that the relationship between spacetime theory and our quantum theories of matter is up for grabs. Everyone involved has to make choices about which assumptions to hold onto and which to let go of as we try to move forward to a unified theory. So these are exactly the moments when stepping back and exploring our assumptions from all sorts of perspectives seems most likely to bear fruit. I haven't engaged directly with that discussion in my paper here, but that's one motivation for wanting to think hard about what's doing what work conceptually. The other is that coming from a philosophy of physics background we tend to take the block universe for granted and to use that as our explanatory basis. *Obviously* there's no such thing as now, *obviously* temporal becoming is in an important sense illusory. But what I've offered you here is an approach from within physics itself that rejects the block universe and opens the door for (i) the "now" of any given system being spatiotemporally limited, and (ii) temporal becoming. I think that's cool and worth exploring.

References

Barbour, J. 1999. *The End of Time.* Weidenfeld and Nicolson, London.

Brading, K. 2011. 'On composite systems: Descartes, Newton, and the law-constitutive approach', in P. Anstey and D. Jalobeanu (ed), Vanishing Matter and the Laws of Nature: Descartes and Beyond, Studies in Seventeenth-Century Philosophy, Routledge, pp. 130–152.

——— 2012. 'Newton's law-constitutive approach to bodies: a response to Descartes', in A. Janiak and E. Schliesser (ed) Interpreting Newton: critical essays (Cambridge University Press). pp. 13–32.

——— 2013 a. 'Three principles of unity in Newton', *Studies in History and Philosophy of Science*, 44, pp. 408–415.

——— 2013 b. 'Physically locating the present: a case of reading physics as a contribution to philosophy', *Studies in History and Philosophy of Science*, 80, pp. 1101-1111.

——— 2013 b. 'Presentism as an empirical hypothesis', *Philosophy of Science.*

Brading and Skiles 'Underdetermination as a path to ontic structural realism', with Alex Skiles, in E. Landry and D. Rickles (ed.), *Structure, Object, and Causality: Proceedings of the Banff Workshop on Structural Realism.* University of Western Ontario Series in Philosophy of Science, Springer, pp. 99–116.

Gibson, I. and Pooley, O., 2006. 'Relativistic Persistence', *Philosophical Perspectives* 20(1), pp. 157–198.

Hinchcliff, M., 2000. 'A defense of presentism in a relativistic setting', *Philosophy of Science* 67 (proceedings), pp. S575–S586.

Putnam, H., 1967. 'Time and physical geometry'. *Journal of Philosophy* 64, pp. 240–247.

Savitt, S., 2000. 'There's no time like the present (in Minkowski spacetime)', *Philosophy of Science* 67 (proceedings), pp. S563–S574.

Stein, Howard. 1968. On Einstein-Minkowski Space-Time. *Journal of Philosophy* 65(1): 5–23.

Van Camp, W., 2011, On kinematic versus dynamic approaches to special relativity, 1097–1107.

A MUSE OF FIRE

JAMES ROBERT BROWN
Philosophy, University of Toronto

1. Introduction

Shakespeare anticipates the work and the wonder of thought experiments in the opening of *Henry V*. He wants us, his audience, to experience the actions of great kings and the clash of mighty armies, but all he has to work with are a handful of puny actors on a barren stage. To achieve the desired result, we must work our "imaginary forces." The passage delivered by the Chorus reminds us how much more than the "willing suspension of disbelief" we must bring to some theatrical situations.

> O for a muse of fire, that would ascend
> The brightest heaven of invention,
> A kingdom for a stage, princes to act,
> And monarchs to behold the swelling scene.
> ...
> ... But pardon, gentles all,
> The flat unraisèd spirits that hath dared,
> On this unworthy scaffold, to bring forth
> So great an object. Can this cockpit hold
> The vasty fields of France? Or may we cram
> Within this wooden O the very casques
> That did affright the air at Agincourt?
> ...
> And let us, ciphers to this great account,
> On your imaginary forces work.
> ...
> Piece out our imperfections with your thoughts:
> Into a thousand parts divide one man,
> And make imaginary puissance.

> Think, when we talk of horses, that you see them
> Printing their proud hooves i'th'receiving earth.
> For 'tis your thoughts that now must deck our kings,
> Carry them here and there, jumping o'er times,
> Turning th'accomplishment of many years
> Into an hourglass...

We are certainly more than passive observers of the passing parade. Most philosophers would acknowledge that. Philosophical disputes arise over what we bring to experience and imagination, and these disputes bear especially on thought experiments. Where do they come from? How do they work? Are they reliable? Can we sort the good ones from the bad? These are some of the questions that arise when we use thought experiments to investigate any subject and we rightly wonder about the legitimacy of that tool. As for the question, How do thought experiments work? there are several answers. The champions of thought experiments may quarrel over the details, but they are united in thinking them an invaluable tool in both science and philosophy. However, there are also detractors. For them the muse of fire would be useful only for an act of arson, to set alight a house of cards, as they see it.

Protesters must be pacified; their objections must be met. Alas, *imagining* the detractors defeated will not make it so. Reasons are required, and so I will play on readers' non-imaginary forces in doing so, calling on their imaginary forces only to illustrate.

2. Challenges to Thought Experiments

Philosophers reflect on thought experiments from both philosophy and science — but philosophers also produce the ones from philosophy, whereas scientists produce the ones in the sciences. This may seem a trite observation, but it could be important in light of recent challenges to the legitimacy of thought experiments. Can we trust our intuitions? I'm not worried about falliblism, but rather about a possible problem with intuition that could be systematic. Philosophical cases may be different from those in physics. In the latter case we have the entire history of science to draw on. Our intuitions (ie, the intuitions of philosophers and historians) are simply not relevant; it is the intuitions of those working scientists that matter, and there is a huge consensus on the legitimacy of those intuitions. This is not the case with philosophical examples, where there is often significant disagreement. This disagreement has played into the hands of critics of thought experiments, critics who stress the fallibility and unreliability of intuition.

Daniel Dennett has used the expression "intuition pump" in a deliberately disparaging way to describe the process of coming to conclusions based on thought experiments. He's done this in a number of publications over several years, but the message is always the same: thought experiments can be fun and even sometimes instructive, but they are not a reliable source of information about anything and they are often highly misleading. We should, he claims, always resort to the empirical study of the mind and nature, rather than rely heavily on any sort of conceptual analysis. Empirical science is where the real surprises lie, and a reliance on thought experiments is nothing more than a reliance on our prejudices.

Dennett has doubtless noticed that thought experiments take a great many forms. The relevant thing about thought experiments is that we can learn something just by thinking — at least, so it seems. Some can be duplicated as real experiments, others perhaps not. I would say that it does not matter — at least not much. We might learn a few empirical details by actually dropping objects from the Leaning Tower of Pisa. We might also get drastically different results, showing that the thought experiment is utterly misconceived. This is what Dennett has in mind. Why perform thought experiments when real ones can do the same or better?

But what about examples like the ship of Theseus? Remember that the parts of the ship are all replaced over time, so that eventually there is nothing left of the original ship. Is the result of this ongoing change still the ship of Theseus? The thought experiment does not answer the question of identity — it only raises the question. Should we heed the critics of thought experiment and stick to real experiments? We can easily duplicate this thought experiment in reality. But it would not help in the least to settle the question of identity. The issue is conceptual, not factual, if I may put it that way, and the conceptual problem is made perfectly manifest by the thought experiment. Nothing could be gained by our understanding by replacing each and every part of a *real* ship.

Dennett is far from alone in pooh-poohing thought experiments. Stephen Stich in recent years has been highly influential in promoting experimental philosophy. He aims to solve philosophical problems empirically, rather than through conceptual analysis. (See Machery, *et al.* 2004) Paul Thagard holds much the same view, putting it rather bluntly: "The currently common use of thought experiments in philosophy is akin to reasoning based on faith rather than evidence." (2010, 14) Both wish to see a great deal more attention paid to the recent results of cognitive science and the neurosciences rather than the results of traditional philosophical techniques.

There is no question that much of their work is interesting and important and it would be foolish to dismiss it out of hand. Some of the consequences are also potentially of social importance. For instance, Stich claims that men and women frequently have different intuitions about important philosophical examples, that is, they do not find the same thought experiments equally plausible. (Buckwalter and Stich, 2010) Stich believes that this is at least part of the explanation for the relative lack of women in philosophy. Everyone's philosophical education involves going through classic thought experiments where one is presumed to have the same intuitions as the community at large and as the instructor, in particular — otherwise one is judged to be simply not cut out for philosophy. Often, Stich finds, women disagree with the standard understanding of some of these classic examples, which leaves them feeling discouraged. Stich claims, they should have no such negative feelings and instead we should all conclude that our different intuitions are simply systematically unreliable. We should abandon thought experiments as a tool and intuition as a source of knowledge. In a similar vein Richard Nisbett (2003) claims that cultural differences matter: Asians think differently than Westerners and come to different conclusions when doing thought experiments. If he were right — which I doubt — it would undermine the confidence we have in thought experiments as a reliable tool.

When we look at the examples cited by Dennett, Stich, Thagard, and Nisbett we find a pattern: they're all drawn from philosophy. Of course, that's hardly surprising, but it is highly important to note this fact. It would be nice to have a single account of thought experiments that covers all examples in philosophy, physics, and every other discipline. But given the challenge to thought experiments, it is important to note that the contested examples are all philosophical, not scientific.

If we look to the history of physics, it is very hard to ignore the fact that thought experiments have played a huge role. There is Galileo on falling bodies, Einstein's elevator, Maxwell's demon, Schrödinger's cat, and so on, as we noted earlier. There is in the physics community a consensus on the importance of these examples and the role they play. Every historian of science would consider them an essential part of the story of how we got to where we are. There is no way they could be reasonably left out. What moral should we draw from this simple fact? We should conclude that thought experiments in physics work and they work rather well. A historian of physics who expressed doubt concerning the significance of thought experiments in the history of physics, the way Dennett and others cast doubt on them in philosophy, would simply be dismissed out of hand as an incompetent historian.

For the moment, let's set aside thought experiments in philosophy. The question then is to make sense of thought experiments in the sciences. Why do they work? But first, how are we to go about answering that question? Well for one thing, we do not have to rely on our intuitions. It is the scientists themselves, Galileo, Einstein, and others, who had intuitions. As historians or epistemologists of science we need to give an account of what is going on and we may do that by positing intuitions — not our intuitions, but Galileo's. We are simply in the business of trying to give the best account we can of the history of science. If intuitions are part of that account, then so much the better. Of course, we may try to replicate a thought experiment just as we often try to replicate a real experiment, and our intuitions in the former case and our observations in the latter will play a role. This can shed light on the past, but only secondarily. It is the thought experiment or real experiment in their historical contest that is crucial.

Now there's an assumption we want to make about the history of science which we would be reluctant to make about the history of philosophy. The assumption is this: *natural science has made significant progress.* If there is such a thing as knowledge (or at least justified belief), it is science that has provided it. We do not know how or what the details are; those are features to be settled in future debate. But if we can be sure of anything, then we should be sure that in some important sense science is a success. If a sceptic were to give us an argument that concluded science has made no progress, we should rightly take that as a reductio ad absurdum of the premises. Most of us would not, however, make a similarly confident claim about the history of philosophy or the social sciences.

This assumption about science gives us something powerful to work with, namely, a subject matter on which we can test our various accounts of thought experiments. The philosophical case is much weaker, and hence more vulnerable to the kind of challenges made by those who have serious doubts about intuition. We do not have an independent testing ground for our thought experiments in the areas of, say, economics or ethics, since so many of the results in these fields are themselves dependent upon thought experiments. And the fact that we do not have a consensus on philosophical thought experiments themselves means that they are doubly unreliable, at least in the view of any one who is already sceptical.

I raised the sceptical doubts of Dennett and others, but pointed out that at best their criticisms applied only to some thought experiments, not to all of them. It may still turn out that intuitions are perfectly legitimate, as well as highly important, in the history of science. That means we need an account of them. But it would also seem that thought experiments in physics have the same structure (or structures, since there are many different kinds) as they have in philosophy. This strongly suggests that whatever account we give, thought experiments ought to have the same status in both fields, and that means, if they are indeed legitimate in physics, then they are also legitimate in philosophy. What we should not assume, however, is that thought experiments will generally be as easy to perform or as reliable in one field as in others. For instance, physics is replete with thought experiments whereas chemistry seems to have none at all. It might also be the case that physics is a natural home for thought experiments, because of its conceptual structure, where a kind of conceptual analysis is possible. Perhaps chemistry is too complex. And perhaps philosophy is somewhere in between, such that we can perform some thought experiments but not with the same clarity as they typically occur in physics. Thus, it would be no surprise if there were a failure of consensus on particular philosophical examples.

Now we face a dilemma: if we side with the critics, then either we cannot adopt a single account of thought experiments or we must abandon the belief that thought experiments have been importantly effective in the history of physics. The alternative is to reject the critics' claim that intuitions are mere prejudices and unreliable sources of knowledge. It will come as no surprise that I recommend the latter view, but I also acknowledge that there is still a challenge from the critics that must be met directly. And this is what I will now do, focussing on Dennett.

3. Dennett's Intuition Pumps

Dennett's influential critiques of thought experiments rest on two related complaints. The first is that they rely on "intuition pumps," an illusory and unreliable source of evidence. The second problem, as he sees it, is that thought experiments rely on folk concepts, which are inherently conservative and unlikely to be the right concepts for the scientific understanding of the world. Let's begin with the first of these complaints.

> I coined the term "intuition pump," and its first use was derogatory. I applied it to John Searle's "Chinese room," which I said was not a proper argument but just an intuition pump. I went on to say that intuition pumps are fine if they're used correctly, but they can also be misused. They're not arguments, they're stories. Instead of having a conclusion, they pump an intuition. They get you to say "Aha! Oh, I get it!" (Dennett 1995)

The problem with them, according to Dennett, is that they give us the illusion of understanding, when in fact they may be nothing more than habits of mind or the cognitively useful products of evolution. Neither is a guarantee of truth. Of course, Dennett is right about the lack of a guarantee; they are not fully reliable. The danger, however, is that we will follow him in throwing out the baby with the bathwater. Qualia, for instance, is something he would like to discard.

Qualia are the subjective aspects of experience, feelings of hunger, pleasure, anger, and sensations of colour, smell, and so on. (One quale, many qualia.) They are accessible to introspection. The status of qualia is central to the mind-body problem. Physicalists claim that there is nothing over and above physical facts, so, qualia present some sort of challenge. Are qualia different? Can they be reduced to the physical, or perhaps eliminated entirely? If not, then physicalism would seem to be wrong.

Frank Jackson is a long-time champion of qualia. He produced a famous thought experiment that has been much discussed for three decades.

> Mary is a brilliant scientist who is, for whatever reason, forced to investigate the world from a black and white room via a black and white television monitor. She specializes in the neurophysiology of vision and acquires, let us suppose, all the physical information there is to obtain about what goes on when we see ripe tomatoes, or the sky, and use terms like 'red', 'blue', and so on. She discovers, for example, just which wave-length combinations from the sky stimulate the retina, and exactly how this produces via the central nervous system the contraction of the vocal chords and expulsion of air from the lungs that results in the uttering of the sentence "The sky is blue." (It can hardly be denied that it is in principle possible to obtain all this physical information from black and white television, otherwise the Open University would of necessity need to use colour television.)
>
> What will happen when Mary is released from her black and white room or is given a colour television monitor? Will she learn anything or not? It seems just obvious that she will learn something about the world and our visual experience of it. But then it is inescapable that her previous knowledge was incomplete. But she had all the physical information. Ergo

> there is more to have than that, and Physicalism is false.
>
> Clearly the same style of Knowledge argument could be deployed for taste, hearing, the bodily sensations and generally speaking for the various mental states which are said to have (as it is variously put) raw feels, phenomenal features or qualia. The conclusion in each case is that the qualia are left out of the physicalist story. And the polemical strength of the Knowledge argument is that it is so hard to deny the central claim that one can have all the physical information without having all the information there is to have. (Jackson 1982, 130)

Dennett outright rejects this thought experiment for the same general reason he rejects all others.

> Like a good thought experiment, its point is immediately evident even to the uninitiated. In fact it is a bad thought experiment, an intuition pump that actually encourages us to misunderstand its premises. (Dennett 1991, 398).

Fortunately, Dennett condescends to play the thought experiment game and he does so with considerable success. Dennett's attack on Jackson is in the form of the following counter thought experiment. The setup is the same as Jackson's, but the subsequent scenario is quite different.

> And so, one day, Mary's captors decided it was time for her to see colours. As a trick, they prepared a bright blue banana to present as her first colour experience ever. Mary took one look and said "Hey! You tried to trick me! Bananas are yellow, but this one is blue!" Her captors were dumbfounded. How did she do it? "Simple," she replied, "you have to remember that I know everything — absolutely everything — that could ever be known about the physical causes and effects of color vision. So of course before you brought the banana in, I had already written down, in exquisite detail, exactly what physical impressions a yellow object or a blue object (or a green object, etc.) would make on my nervous system. So I already knew exactly what thoughts I would have (because, after all, the mere disposition to think about this or that is not one of your famous qualia, is it?). I was not in the slightest surprised by my experience of blue (what surprised me was that you would try such as second-rate trick on me). I realise that it is hard for you to imagine that I could know so much about my reactive dispositions that the way blue affected me came as no surprise. Of course it's hard for you to imagine. It's hard for anyone to imagine the consequences of someone knowing absolutely everything physical about anything! (Dennett 1991, 399f)

It should be quite clear at this point what is happening. Jackson's thought experiment has the following structure: There is a set up: Mary is in a black and white room, where she learns all physical facts about perception having only black and white experiences. Next comes the phenomenon: When Mary first encounters colours, she learns something new. Finally, the result of the thought experiment, i.e., the conclusion Jackson draws: Some facts about perception are not physical, and so, physicalism is wrong. *Counter thought experiments*, as I shall use the term, accept the set up, but challenge the phenomenon, which is exactly what Dennett does. This is an unusual way to reason, unlike the normal challenge to an argument. Usually, when we reject a conclusion, we either blame a premiss in the argument or we deny that the conclusion follows from the premisses. In short, we challenge soundness or validity. In a thought experiment, there is an intermediate stage which is also open (in principle) to challenge — the phenomenon. We set things up (similar to adopting some premisses), then let it run; we see what happens (this is the phenomenon, which we observe in the thought experiment), then we draw a conclusion (validity is at issue at the last step). A counter thought experiment is not merely a thought experiment that comes to a rival conclusion. It is (as defined here) a thought experiment with the same set up (same initial premisses), but the subsequent unfolding narrative presents a very different phenomenon. Unsurprisingly, a different conclusion is then drawn.

The example shows Dennett to be a clever thought experimenter, indeed, a clever counter thought experimenter. We might point out that he seems to be somewhat inconsistent in using such a technique, having just denounced it in no uncertain terms. That would be unfair. He need not embrace thought experiments, he need only point out that either thought experiments work or they do not. If they do not, then Jackson's argument is undermined, since Jackson uses a thought experiment to get his result. If they do work, then Jackson's result is undermined by the counter thought experiment. One way or the other, Jackson's result is undermined.

4. Dennett and Conservative (Folk) Concepts

Dennett offers a brief account of what is really going on generally in thought experiments and why it's a problem. This is the basis of his second main complaint.

> Thought experiments depend on folk concepts; they are inherently conservative. We should expect very counter intuitive results in real science, so violating our intuitions is to be expected. (Dennett 2005, 128f)

He is at least half right. Many of the concepts that play a role in thought experiments are intuitive and commonsensical. Relying on them is indeed a conservative thing to do. Many of the concepts that will be needed to understand the world will be like nothing we currently grasp. If thought experimenting stood in the way of acquiring them, we would be in serious trouble. Sometimes, however, we have a cluster of intuitive, commonsensical concepts that either have not been fully fleshed out or might even be in conflict with one another. Thought experiments can often make great advances here.

Since the centennial in 2012 of Alan Turing's birth has just passed, it seems particularly appropriate that we use his analysis of computation to show how Dennett's claims are mistaken. Turning's analysis of computation (Turing 1936) could legitimately be called a thought experiment, but that is not important. What matters as far as our case against Dennett goes, is that intuitive pieces are pasted together into a very surprising result. It takes the form of a simulation of what a person does when making a computation or following some algorithmic procedure. We know what it is like to calculate something such as the product of two numbers that are multiplied together. Turning imagines an analogue to this is the form of a machine.

> We may compare a man in the process of computing a real number to a machine which is only capable of a finite number of conditions $q_1; q_2; \dots, q_n$ which will be called "m-configurations". The machine is supplied with a "tape" (the analogue of paper) running through it, and divided into sections (called "squares") each capable of bearing a "symbol". t any moment there is just one square, say the r-th, bearing the symbol $\mathfrak{S}(r)$ which is "in the machine". We may call this square the "scanned square". The symbol on the scanned square may be called the "scanned symbol". The "scanned symbol" is the only one of which the machine is, so to speak, "directly aware". However, by altering its m-configuration the machine can effectively remember some of the symbols which it has "seen" (scanned) previously. The possible behaviour of the machine at any moment is determined by the m-configuration q_n and the scanned symbol $\mathfrak{S}(r)$. This pair q_n, $\mathfrak{S}(r)$ will be called the "configuration": thus the configuration determines the possible behaviour of the machine. In some of the configurations in which the scanned square is blank (i.e. bears no symbol) the machine writes down a new symbol on the scanned square: in other configurations it erases the scanned symbol. The machine may also change the square which is being scanned, but only by shifting it one place to right or left. In addition to any of these operations the m-configuration may be changed. Some of the symbols written down will form the sequence of figures which is the decimal of the real number which is being computed. The others are just rough notes to "assist the memory". It will only be these rough notes which will be liable to erasure. It is my contention that these operations include all those which are used in the computation of a number. The defence of this contention will be easier when the theory of the machines is familiar to the reader. In the next section I therefore proceed with the development of the theory and assume that it is understood what is meant by "machine", "tape", "scanned", etc. (Turing 1936–7, 231)

The background to this is worth noting. Hilbert was interested in finding a mechanical procedure that would allow us to settle every mathematical question. It was known as the *Entscheidungsproblem*, the decision problem. We have methods for doing long division, calculating square roots, and so on. Perhaps there is some general procedure that could be applied to any problem, a method that would mechanically lead us to the correct result. Or, at least, for any problem, there is a mechanical method to solve it. Hilbert and others were optimistic that such procedures could be found; others were not. Turing settled the issue in the negative: no such procedure exists. To show this, he had to say just what an algorithm or mechanical procedure is. His answer, now known as a Turing machine, is generally accepted as the correct analysis.

As we work through Turing's analysis, keep Dennett's claim in mind. Recall the passage quoted above, which I now repeat: "Thought experiments depend on folk concepts; they are inherently conservative. We should expect very counter intuitive results in real science, so violating our intuitions is to be expected." As we shall see, we can grant the first claim — thought experiments depend on folk concepts — and still agree with the second claim that real science will be counter intuitive, even when we arrive at that result with a thought experiment.

Turing, as we saw in the quoted passage, employs the idea of an indefinitely long tape with squares on that in which could be written a 0 or a 1. Suppose it is everywhere 0, except a finite number of squares (possibly none), with a 1 on them. He imagined a simple machine with a head that could read and write and could move to the left or the right. The specific operations include: "move left one square," "move right one square," "erase what is on the square and write 0," "erase what is on the square and write 1." The machine (a Turing machine), could be given a set of instructions to follow. They are in conditional form, such as: "if in state S_3 reading a 0, then write a 1, move left one square and go to state S_5", "if in state S_6 reading a 0, then write a 0, move right one square, and go to state S_2", and so on.

How is this intuitive or commonsensical? Notice how it corresponds to what we do when we calculate. We have a sheet of paper, which keeps track of the stages of a calculation. This is the memory and it corresponds to the tape in the machine. We write symbols on the paper; the machine writes 0s and 1s on the tape. We write those symbols in an orderly way on the page; the machine writes linearly on the tape. When working through an algorithm for addition, we might first add the tens, then we add the hundreds, and so on. We are in a different mental state in the various stages. Similarly, the machine will be in one state and will perform some action based on being in that state and reading some input on the tape.

A finite list of such instructions is a programme, software. It can be put to work on some input, that is, on a tape with a finite string of 1s, and halt after generating some output, another string of 1s. For instance, the initial tape might be: ...000110111000... and the final tape is ...00011111000... The programme that generated this would be doing the computation of $2 + 3 = 5$.

After constructing a very wide range of examples, people became convinced that any function that seemed intuitively computable could be computed by a Turing machine. This gave rise to a conjecture that is now more or less universally accepted, the Church-Turing Thesis: *Every computable function can be computed by a Turing machine.*

There is no denying that this is a stupendous achievement. For centuries we have had an vague understanding of algorithmic processes. Thus, as school children we all learn how to do long division; we carry out a definite procedure and we arrive at the correct answer. But for other things, such as finding the roots of a fifth degree polynomial, we needed guess work. Thanks to Turing, we now have a definitive analysis of what an algorithm or a mechanical procedure is. But this is just the beginning. Turing was able to prove remarkable things thanks to his thought experiment, that is, his analysis of intuitive or folk concepts. For instance, he established that most functions are not computable. The proof is easy: Since any Turing machine programme is a finite list of instructions, there are at most countably many distinct ones. The set of functions, like the set of real numbers, is an uncountable infinity. Thus, most functions are uncomputable.

These are spectacular results. The Church-Turing thesis is based on experience and the uncomputablity of most functions is based on transfinite set theory. Neither are thought experiments themselves. But they are based on the analysis of computation, that is, on the idea of a Turing machine, and that is all part of a thought experiment or something close to it. We could readily agree with Dennett that each of the concepts involved, "erasing a 0," "moving right one square," and so on, is intuitive, commonsensical, an utterly conservative folk concept. There is nothing new here in any of these notions. But the mixture is new and the results are anything but intuitive, commonsensical, and conservative.

I said Turing's analysis of computation might be a thought experiment. Maybe it was or maybe it wasn't — it does not matter. The point I am trying to emphasize is that we can start with commonsense concepts and finish with something highly surprising. Of course, lots of great science is not like this at all. Concepts such as the electromagnetic field or germs as the causes of disease were novel when first introduced. Presumably Dennett has this sort of thing in mind when he tells us to expect our old beliefs to be overthrown. He is right, of course, about much of science, but it is not the whole story. Some stellar and surprising achievements are based on what is ready at hand. Some remarkable achievements result from re-jigging everyday items and bringing their consequences to light. This is often how thought experiments provide new breakthroughs.

5. Weird Spacetimes

There is much more to be said on this topic. Dennett's claims about conservative folk concepts are upended by Turing machines. Now let's mix Turing's analysis with different spacetimes. Needless to say, many of the main concepts of General Relativity are not conservative folk concepts by anyone's lights.

It turns out that what is and what is not computable does not rest solely with what the Turing machine can or cannot do. It rests in part on the physics of the world in which a computation is carried out. Consider the following remarkable example. It is similar to the twin paradox, except that one twin passes a finite time while the other passes an infinite time.

Near a black hole two observers γ_1 and γ_2 move along future-directed time-like curves in such a way that γ_1's curve is infinite while γ_2's is finite. Such spacetimes have been constructed by Pitowsky (1990) and by Hogarth (1994) and Malament.

Suppose γ_1 can send signals to γ_2. γ_1 has a Turing machine and is checking, say, Goldbach's conjecture, working through the numbers, one after the other. For simplicity, we may consider γ_1 to be a Turing machine.

If γ_1 finds a counterexample to Goldbach's conjecture (which would happen at some finite time in γ_1's infinite proper time, if a counterexample exists), then a signal will be sent from γ_1 to γ_2.

If γ_2 does not receive a signal by the time some spacetime point p is reached, then γ_2 will know that there is no counterexample.

The upshot of this thought experiment is profound. What is computable depends (in part) on the nature of the physical universe in which the computation is done. In a Newtonian universe, such a computation is not possible. But in a spacetime allowed by General Relativity, near a black hole, it turns out that Goldbach's conjecture is computable. The theory of computation is not pure mathematics, but at least in part physics. This discovery is in part the result of an analysis of conservative folk concepts, the very thing that Dennett thinks is a mere reflection of prejudices we could better do without. The opposite is true.

Bibliography

Brown, J. R. (1986) "Thought Experiments Since the Scientific Revolution," *International Studies in the Philosophy of Science*, Vol I, no 1. 1986.

Brown, J. R. (1991/2010) *Laboratory of the Mind: Thought Experiments in the Natural Sciences*, London: Routledge.

Brown, J. R. (1999/2008) *Philosophy of Mathematics: An Introduction to the World of Proofs and Pictures*, London: Routledge.

Buckwalter, Wesley and Stephen Stich, (2010) "Gender and Philosophical Intuition" (September 26, 2010). Available at SSRN: http://ssrn.com/abstract=1683066 or http://dx.doi.org/10.2139/ssrn.1683066.

Dennett, D. (1991) *Consciousness Explained*, New York: Little Brown.

Dennett, D. (2005) *Sweet Dreams*, Cambridge, MA: MIT Press.

Dennett, D (1995) "Intuition Pumps" in J. Brockman (ed.) *The Third Culture*, New York: Simon and Schuster (also available in part: www.edge.org/documents/ThirdCulture/r-Ch.10.html).

Hogarth, M. (1992) "Does general relativity allow an observer to view an eternity in a finite time?" *Foundations of Physics Letters*, 5(2):173–181.

Hogarth, M. (1994) "Non-Turing computers and non-Turing computability," PSA: *Proceedings of the Biennial Meeting of the Philosophy of Science Association*, Vol. 1, 1994:126–138.

Hogarth, M. (2004) Deciding arithmetic using SAD computers. British Journal for the Philosophy of Science, 55:681–691.

Machery, Edouard, Ron Mallon, Shaun Nichols, Stephen P. Stich (2004) "Semantics, cross-cultural style," *Cognition*, 92 (2004) B1–B12.

Nisbett, Richard (2003) *The Geography of Thought: How Asians and Westerners Think Differently... and Why*, Free Press: New York.

Pitowsky, I. (1990) "The physical Church-Turing thesis and physical computational complexity," *Iyyun*, 39:81–99.

Thagard, Paul (2010) *The Brain and the Meaning of Life*, Princeton: Princeton University Press.

Turing, A.M., 1936–7, "On Computable Numbers, With an Application to the Entscheidungsproblem," *Proceedings of the London Mathematical Society*, s2–42: 230–265; correction *ibid.*, s2–43 (1936):544–546 (1937).

SSK OR ESW? — THE BLOOR-LYNCH DEBATE REVISITED

KAI-YUAN CHENG

Institute of Philosophy of Mind and Cognition,
National Yang-Ming University

Philosophical discussions of rule-following in the later Wittgenstein (1953, 1967) are an important source of inspiration for the development of views on the social nature of scientific knowledge. Two major opposing views in this inquiry — Bloor's sociology of scientific knowledge (SSK) (1983, 1991, 1992, 1997) and Lynch's (1992, 1993) ethnomethodological studies of work (ESW) — represent two positions derived from two different readings of Wittgenstein's later writings on rule-following. The aim of this paper is two-fold. One is to re-examine the noted Bloor-Lynch debate by considering Kusch's (2004) recent discussion of this debate. Another is to show that a new semantic framework of rule-following ascriptions based on a cognitive approach to the study of generics can be provided such that SSK and ESW are compatible in it (Leslie, 2009; Cheng, 2011).

I. Introduction

Wittgenstein's (1953, 1967) later writings are undoubtedly an important source of conceptual resources and inspirations for those who accept and develop the view that scientific knowledge has a social dimension in essence. Kuhn (1962), for instance, exploits the notion of family resemblance to illustrate the idea of paradigm which figures centrally in his explanation of scientific knowledge. In Kuhn's view, members of a scientific community during the stage of normal science acquire and transfer their competences through researching and operating on a set of exemplars which appear in textbooks, classrooms and experiments. These exemplars constitute a paradigm. A paradigm, however, is grasped without there being a rule of some sort underlying those exemplars. Similar to the fact that some members of a family may resemble some other members of a same family without there being some common feature shared by each member, exemplars do not constitute a paradigm by possessing a same characteristic. Thus, a paradigm is always prior to a rule, in Kuhn's idea. Since a paradigm is determined not by an independently existing rule, but by implicit practices and consensuses of the members of a scientific community, the ways in which normal sciences operate obviously have a social aspect.

Compared to Kuhn, Bloor (1976, 1983, 1997) — a representative advocate of the strong program in the *Science and Technology Studies* (STS) — has relied on the later Wittgenstein's philosophy in a more substantial and in-depth manner to show that scientific knowledge is social in nature. Bloor focuses his inquiry on the notions of a rule and of what it is to follow a rule. These notions occupy the later

Wittgenstein's central concerns, and obviously touch on a similar matter as that of family resemblance, but go deeper than that. The basic strategy of Bloor's argument for SSK is this. In light of the facts that the issue of rule-following deals with the most fundamental aspect of language — what it is for a symbol to have content or meaning, and that mathematics and sciences cannot but express and deliver contents through symbols of various types, an answer to the question of what it is to follow a rule will be critical to illuminate the nature of scientific and mathematic knowledge. If language is as Wittgenstein maintains social in nature, it naturally follows that scientific knowledge is social in essence.

The recourse to Wittgenstein's philosophy in the task of uncovering the nature of scientific knowledge is certainly fruitful, but not at all straightforward. Wittgenstein's philosophical writings are notably obscure; they do not appear to contain arguments, consisting of explicitly stated premises and conclusions, as typically seen in most philosophical works. This renders reading Wittgenstein difficult and controversial. It is in this context that Lynch's (1992, 1993) work is relevant and important. Lynch opposes to Bloor's SSK not only by giving an interpretation of Wittgenstein's writings on the topic of rule-following that he deems more correct, but also by offering a different methodology of doing sociological studies of scientific knowledge based on his interpretation. The dispute between these two positions is widely known as "the Bloor-Lynch debate".

The purpose of this paper is to re-examine this debate in light of subsequent discussions. The plan is as follows. After introducing this debate (section III), I shall discuss Kusch's (2004) recent criticisms of these two interpretations, which lead to the apparent instability of SSK and ESW (section IV). I then attempt to show that i) a generic semantic account of rule-following ascriptions (Cheng, 2011; Leslie, 2009), which is meanwhile a plausible construal of Wittgenstein's view, can be provided, and that ii) this account may serve as a semantic framework where the legitimacy and comparability of SSK and ESW can be granted (section V). But, first, in section II, we will start with Kripke's reading of Wittgenstein, which is highly influential when it comes to getting a clear picture of what the later Wittgenstein tries to say about rule-following.

II. The Rule-Following Problem in Kripke's Wittgenstein

Kripke's reading of Wittgenstein (KW) regarding the problem of rule-following is commonly consulted and compared by both sides of the Bloor-Lynch debate for the purpose of illustrating their positions on the social nature of scientific knowledge. This is not surprising, given KW's penetrating clarity. KW consists of two claims: 1) the rule-following passages embody a sceptical argument leading to a paradox that no fact of the matter can determine whether one follows a rule or another; 2) the paradox can be resolved by a sceptical solution which amounts to suggesting that rule-following ascriptions remain perfectly meaningful even though those ascriptions correspond to no facts. It will be useful to briefly review each claim, respectively.

KW presents us with a sceptic who challenges me why "125" — an answer that I give with confidence after careful calculation — rather than "5" is the correct answer to give to a query "68 + 57" that involves numbers so big that I have never encountered before. The point of the challenge is that if I have been following the *plus* rule or using the "+" sign to mean *addition*, I should now respond "125" to the query "68 + 57". However, if I have been all along following the *quus* rule or using the "+" sign to mean *quaddition*, I should respond "5". The *quaddition* function coincides with the *addition* function for cases of x + y = z where x or y are less than 57. Otherwise the *quaddition* function gives the result of 5. The challenge then amounts to citing a fact about me that I have been following the *plus* rule rather than the *quus* rule.

KW's sceptic proceeds with a method of elimination: several of the most promising candidate facts that may constitute rule-following are examined one by one, but each is found to be either inadequate or inappropriate. It is important that an appropriate fact that constitutes rule-following cannot already contains intentional or semantic content, for it will amount to a circular account and render the task of meeting the sceptic's challenge trivial. Proposals about rule-following constituting facts such as my past usage, my disposition, my mental image or qualia, a formulation that occurs in my mind or verbal articulations, Platonic entities that I "grasp", etc., run into difficulties of one form or other, and consequently, none of them is sufficient to determine whether I meant *addition* or *quaddition* by the "+" sign. This is the sceptical paradox about rule-following concluded by KW.

The sceptical solution accepts this conclusion, but nonetheless tries to remedy its highly counterintuitive and disastrous consequences by showing that our practice of ascribing rule-following to people remains legitimate. The trick is to replace the truth-condition semantics with a form of assertability-condition semantics. The latter view has it that the meaning of a rule-following ascription such as "John means *plus* by '+'" or "Mary follows the *plus* rule" is given by a set of conditions under which such an ascription is assertable, and that the job of delivering a set of assertability conditions is a purely descriptive matter. KW maintains that a close look at how meaning or rule-following ascriptions are actually made reveals that assertability conditions are those where the ascriptions in questions are approved by the majority of members in a community. Insofar as most of the competent adders in a community concur on John's arithmetic performances and are ready to accept him as one of them, the ascription "John means plus by '+'" is assertable, meaningful and true. Since assertability conditions inevitably involve a community, there is an important sense in which ascriptions of rule-following are essentially social.

Despite its clarify and ingenuity, KW's reading has been criticized as giving a false reading of Wittgenstein. McDowell (1984, 1992), for example, cites section 201 to show that Wittgenstein explicitly affirms the reality of rule-following:

> This was our paradox: no course of action could be determined by a rule, because every course of action can be made out to accord with the rule.

> The answer was: if everything can be made out to accord with the rule, then it can also be made out to conflict with it. And so there would be neither accord nor conflict here.
>
> It can be seen that there is a misunderstanding here from the mere fact that in the course of our argument we give on interpretation after another; as if each one contented us at least for a moment, until we thought of yet another standing behind it. What this shows is that there is a way of grasping a rule which is *not* an *interpretation*, but which is exhibited in what we call "obeying the rule" and "going against it" in actual cases (Wittgenstein, 1953, section 201).

As McDowell understands it, the purpose underlying Wittgenstein's engaging in skeptical considerations about various rule-constituting facts is not to cast the reality of rule-following into doubt, but to expose the problematic assumption of reductionism implicitly adopted in those skeptical considerations. In the view of McDowell's Wittgenstein (MW), many vigorous positions that try to explicate rule-following by invoking certain more basic notions tend to oversimplify the phenomenon under scrutiny and ends up in giving a misguided depiction of it. MW opts for a quietist meta-philosophical position — metaphysical debates are impossible and in vain — to deal with the problem of rule-following. In this view, there are no intention- or content-free vocabularies to explain the phenomenon of rule-following. The best we can do is to notice that the phenomenon in question is already out there, embedded in our everyday practice, ready to be observed in we tune our perspective appropriately.

It might be contested that a false reading does not mean the same thing as an incorrect view. Perhaps KW's position, though not faithful to Wittgenstein's own, is convincing, and that is all we need. The prospect of taking this route seems, unfortunately, dim. Quite a few philosophers have quite compellingly pointed out that KW's position is internally incoherent or unstable (see, e.g., Blackburn, 1984; Hoffman, 1985; Boghossian, 1989; etc.). It is argued that once the skeptical conclusion is accepted, the utterances of members of a community which express agreement and disagreement among those members will be void of content. This renders the skeptical solution a hopeless move.

Bearing KW's power and defects in mind, we may proceed to see whether and how Bloor and Lynch may do better than Kripke in reading Wittgenstein.

III. The Bloor-Lynch Debate: SSK or ESW

Bloor's (1976, 1983) earlier readings of Wittgenstein's writings on rule-following appear to come close to that of KW. Bloor maintains that Wittgenstein's major contribution in the inquiry of the nature of mathematical rules is the demonstration of the incapability of a rule to determine its correct applications. Take mathematical proof for example. Traditionally, it is held to be the case that objective mathematical rules and facts exist. Hence, a mathematical proof is a process to "discover" those facts. In opposition to this view, Wittgenstein proposes that a mathematical proof be more appropriately described as a process to "invent" facts about theorems and rules. In Bloor's construal, Wittgenstein's intent in the critical examination of mathematical symbols is to uncover a general crisis that can be extended to symbols of our ordinary languages: there is a gap between a symbol and its applications, but we do not seem able to find suitable items to fill the gap.

Lynch (1992), when discussing Bloor's SSK, notices the similarity between KW and Bloor's reading of Wittgenstein, as the following remark shows:

> Bloor (1976, 1983) believes that rules on their own are unable to determine and explain scientists' actions and beliefs. This is clearest in mathematics: mathematical rules on their own are unable to determine mathematical calculations and derivations. For Bloor, we can see that scientists' actions and beliefs are determined, and can be explained, only once we focus on the causal contribution of social factors like interests, conventions or processes of socialization. SSK identifies these social factors and offers causal explanations of scientists' actions and beliefs. SSK is ultimately based upon a certain view concerning rules. This view Bloor finds in Wittgenstein. Bloor interprets Wittgenstein's cryptic remarks on rule-following as presenting us with a 'rule-sceptical' account. According to this account, no rule (formulation) can ever determine how the rule is to be applied.

Lynch's objection to SSK is thus based on its relying on a position which is vulnerable to difficulties similarly faced by KW. A main difficulty, as shown earlier, is that once the sceptical paradox were accepted, the sceptical solution could not do the job as it was promised to. Lynch's (1992) criticizes Bloor below:

> In his book *Wittgenstein on Rules and Private Language* (1982), Saul Kripke has outlined a version of rule-scepticism that is very close to Bloor's proposal. Unfortunately for Bloor, Kripke's book has been refuted — both as an interpretation of Wittgenstein and as a viewpoint in its own right. The decisive criticism was put forward by Gordon Baker and Peter Hacker (1984). They have shown that one cannot understand a rule without knowing how to apply it. Hence rule applications are not underdetermined by rule formulations, and a fortiori, there is no explanatory gap that needs to be filled with social interests and the like. In light of this analysis, we have good reason to give up on causal SSK and turn to ethnomendology. Ethnomethodology of science confines itself to describing scientists' actions.

However, as Kusch (2004) points out, Bloor (1997) has later published a book-length discussion of the rule-following considerations where he rejects KW's rule-scepticism. It would, therefore, be unfair to Bloor if his more updated position were not considered.

What is new in Bloor's (1997) subsequent work is his proposing a theory of social institutions as an account of rule-following in response to KW's skeptical challenge.

This theory is based on a self-referential model of social institutions. Its basic idea is this. Take money for example. What makes a piece of paper cash, or an instance of money? Clearly, it is by virtue of a community's treating, talking of, and believing a piece of paper as cash that makes it so. It is the collective acts of calling and dealing a piece of paper as an instance of money that function as enabling a piece of paper to possess the property of money. The acts in question are self-referring because their invoking of the category "money" creates money as the referent for such acts. Money is a social status imposed by us through our collective acts. In a similar vein, the fact that a response to a query in a rule-following scenario can be correct or incorrect is determined by our calling it "correct" or "incorrect". A rule is rooted in our self-referring practice of deeming a response as conforming to a rule, and of citing and invoking a rule in the course of abiding by it. As Bloor says: "Between reference to an independent reality, and having no reference at all, we have self-reference, i.e., reference to a reality, but a reality which is dependent on the very acts of reference that are directed at it" (1997, p. 68). This is a straight solution to KW's scpetical paradox, and "can be considered as a form of community-wide dispositional theory" (1997, p. 68).

Bloor calls this theory "meaning finitism". Its main idea is that community-wide dispositions to follow a rule are grounded in ostensive training, where instances presented to a trainee in a setting of learning are finite. A complex set of factors ranging from an individual person's natural instincts and biological mechanisms to psychological inclinations comes into play in shaping the rule-following dispositions in a communal setting. Bloor makes it explicit that he aims to supplement Wittgenstein with an empirical program such as SSK of uncovering a rich and complex set of biological, psychological and social factors that enter into the constitution of our mathematical and scientific knowledge. Bloor sees himself as offering a straight solution to the rule-following problem, and does not mind if taking this theory-building approach deviates from what Wittgenstein would endorse. By doing so, he fully acknowledges Wittgenstein's insight but extends beyond it.

Lynch's position stands in sharp contrast with that of Bloor which aims at giving a constructive account. Lynch's intent is to give a faithful reading of Wittgenstein, with the conviction that this work amounts to delivering a compelling view of rule-following. A better and more correct methodology of doing sociological studies of scientific knowledge would can then be based on that.

Lynch's preferred reading of Wittgenstein is derived from the interpretations of several philosophers such as Shanker (1987), Malcolm (1989), and especially Baker & Hacker (1984). What these philosophers' readings share in common is the emphasis of an internal relation between a rule and its applications. Baker & Hacker, for example, contend that KW goes astray at the very beginning of presenting the skeptical challenge. The mistake lies in thinking that there is a third item to be filled in between a rule and its applications such that the determination of correct applications by a rule can be assured. Baker & Hacker argue that only an external relation between two items can admit of further explications. An internal relation is a conceptual one, where the two items in the relation are necessarily linked. For example, the relation between the notion of father and that of son is a conceptual one; someone who is called "father" necessarily has at least a son, and someone who is called "son" must have a father. The two items linked conceptually are necessarily connected. Similarly, in the view of Lynch's Wittgenstein, the connection between a rule grasped by a person and the person's calculations performed in light of the rule is internal and hence conceptual. Apart from the addition rule, there would be no instances of $2 + 2 = 4$ or $23 + 145 = 168$, etc. The reverse is the same, that is, apart from instances of $2+2 = 4$ or $23+145 = 168$, etc., the content of the addition rule would be determined. Thus, when we say that someone grasps some rule, we are already linking conceptually and necessarily the rule to its relevant instances. In other words, there is no need for a rule and its instances to be further determined, as KW's sceptic assumes.

Backed up by this reading, Lynch opposes to both individualistic and communal forms of dispositionalism about rule-following. Both versions wrongly identify the relation between a rule and its applications as external, and end up seeking some constructive, causal account of the relation in question in vain.

Note that this position offered by Lynch's Wittgenstein differs from the skeptical solution provided by KW. The former is diagnostic in nature, unlike the latter embodying a reductionist assumption which encourages giving constructive accounts. Viewing from this non-scepticist perspective, Lynch suggests that Bloor's analytic social science is entirely at odds with Wittgenstein's insight, which emphasizes a non-constructive approach to tackling a philosophical problem. Since the methodology of SSK is of an analytic and constructive style, it is doomed to face severe difficulties raised in Wittgenstein's rule-following considerations. In contrast, as Lynch continues to stress, his reading of Wittgenstein offers a perfect conceptual ground for developing the methodology of ESK. For, as a methodology in the sociological study of scientific knowledge exemplified in the writings of Garfinkel (1967), Garfinkel & Sacks (1970), etc., ESW aims to offer pure descriptions of what scientists do and what scientific practices look like. No constructive work is involved, and hence no risk of running into philosophical challenges is taken.

Here we need to stress the existence of significant differences between the two methodologies adopted by SSK and ESW, for it may appear that those two methodologies are more similar than what the two contenders maintain. After all, proponents of SSK would agree with the assertion of ESW that a sociological inquiry of scientific knowledge requires delicate and comprehensive descriptions and observations of scientific practices. I think the most important difference lies in that, for ESW, a sociological study of scientific knowledge can only observe and describe scientific practices, whereas for SSK, merely observing and describing scientific practices are not enough. The latter wants more: some causal explanation of why and how those observed phenomena take place is required in an adequate sociological inquiry of sciences.

At this juncture, we wonder: which reading is more convincing, and which methodology more plausible? How shall we evaluate them?

IV. Kusch's Criticisms of Bloor and Lynch

Kusch is unsatisfied with neither of the two positions taken by Bloor and Lynch, respectively. As Kusch sees it, Bloor's community-wide dispositional theory is ambiguous regarding whether it gives a reductive or non-reductive characterization of a disposition. This issue matters, because only a reductive dispositional theory can meet the skeptical challenge. A non-reductive account contains intentional or semantic content, and leads to circularity and triviality. Kusch (2004, p. 580) maintains that Bloor's dispositional theory is a non-reductive one, for Bloor extensively employs vocabularies such as "calling institutional meaning facts actors' categories" or "the rule exists in and through the practice of citing it and invoking it" that are obviously intentional social facts.

In response to Kusch, Bloor (2004) stresses that his theory qualifies as a reductive account, given that it concerns the grounding of intentional processes on non-intentional ones. As Bloor explains, "...acts of self-reference (that is, reactions, made within the institutions) are themselves grounded in more basic (non-intentional) processes...the social fact is meant to be analyzed right down to its non-intentional components..." (2004, p. 596). Bloor continues to cite his own writings to defend this construal:

> The real sources of constraints...as we move from case to case are the local circumstances impinging upon us: our instincts, our biological nature, our sense experience, our interactions with other people, our immediate purposes, our training, our anticipation of an response to sanctions, and so on through the whole gamut of causes, starting with the psychological and ending with the sociological. This is the message of Wittgenstein's meaning finitism. (Bloor, 1997: 20).

Kusch does not find Bloor's response convincing, for he thinks that "x is grounded in y" is not equivalent to "x is reducible to y" (2004: 616–7). No one would deny that rule-following as biological, psychological and collective aspects, as Kusch maintains. However, it is a distinct thing to say that rule-following can be reduced to something biological or naturalistic. Even if we grant the construing of Bloor's community-wide dispositional theory as reductive, this theory will be subject to the skeptical challenge. For, all the biological, instinctive and natural psychological proclivities are dispositional in nature. It is unclear whether Bloor is ready and able to counteract the powerful arguments against dispositionalism presented by KW's sceptic to which Bloor's reductive dispositionalism is vulnerable.

Kusch also finds fault with Lynch's non-scepticist position by showing how its basis, e.g., Baker & Hacker's diagnostic reading of Wittgenstein, founders. He offers two major considerations. One is that KW appears to regard the relation between a rule and its applications as internal, not external. Kusch cites the following passage from KW:

> Wittgenstein's view that the relation between the desire (expectation, and so on) and its object must be 'internal', not 'external', parallels corresponding morals drawn about meaning in my text later (the relation of meaning and intention to future action is 'normative', not descriptive...') (Kripke, 1982: 25–6; c.f., Kusch, 2004: 584–6).

Another is that Baker & Hacker's point about the internal relation can tackle at best the epistemological problem of how I know whether a particular response of mine is a correct application of a rule, but not the ontological problem of whether I follow a rule at all. The reason is that an internal relation between x and y brings with it a conceptual and criterial link between x and y. Invoking such a relation can answer epistemic questions about whether an instance of x is y, but not whether there is any fact about the relation between x and y. But clearly, KW's rule-following skepticism is ontological rather than epistemological (Kusch, 2004, p. 585).

To bypass the drawback of Baker & Hacker's reading, Lynch may adopt a similar interpretation of Wittgenstein offered by McDowell (1992), where quietism is articulated and highlighted. This quietist approach emphasizes Wittgenstein's meta-philosophical attitude toward a philosophical problem: a philosophical problem arises as a result of our being misled or misguided by our misuse of language, and thus a right way to deal with it is to dissolve it by bringing a certain way of looking at language into view, rather than to solve it by offering a constructive account of some sort. In this view, the phenomenon of rule-following is as natural as eating and drinking; its essence is laid out as daily life activities open to our eyes for detailed descriptions, with intention and content already inherent in it. There is no need to reduce rule-following to some other more fundamental items.

Putting aside whether McDowell's Wittgenstein (MW) is the real Wittgenstein, and whether it is an obtainable position, it is unclear whether a quietist line of thought can lend support to the methodology of ESW. A major reason is that MW's quietism may be geared toward the treatment of a philosophical problem, not a problem in social sciences. That is to say, MW may well think that social scientific practices have their own "forms of life", have developed methodologies and practices that emphasize seeking causal explanations. As a result, MW may respect practices that try to dig deeper into what lies at the surfaces. In other words, if MW's quietist approach is confined to tacking a philosophical problem, it will support SSK, not necessarily ESW. And this would not be a result welcomed by Lynch.

If Kusch's criticisms of Bloor and Lynch are effective, both SSK and ESW will be endangered. Kusch nonetheless claims that Bloor's SSK is not threatened by KW. The reason is that though Bloor's position seeks to explain mathematical and scientific knowledge in terms of social entities and is hence reductive, its causal explanations remain in the intentional realm, for the notions of action, beliefs and desires are kept in those explanations (Kusch, 2004: 587).

It is unclear whether Kusch's defense of Bloor's SSK works, for two things. One is: isn't Bloor's non-reductive dispositionalism subject to the charge of circularity raised by KW, and the effectiveness of the charge is acknowledged by Kusch? Another is: has Bloor a ready and plausible response to how to defend his position against KW's skeptical argument against dispositionalism in general, especially the finitude and normativity problems (see Kripke, 1982: Chapter II)? If not, then it is hard to see how SSK can be saved from the threat of KW.

We now face the following dilemma. One the one hand, if we opt for Bloor's SSK, it is based on a self-referential model and, as a dispositional account of some sort, falls prey to KW's skeptical challenges. On the other, if we adopt Lynch's ESW, the Baker-Hacker reading of Wittgenstein on which it relies seems insufficient to respond to KW's metaphysical rather than epistemological skeptical challenges. Even if a similar route were taken, i.e., McDowell's quietist reading of Wittgenstein, uncertainty about its applicability to social scientific practices cannot be precluded. Therefore, our difficult situation seems to be: do we need to abandon both SSK and ESW?

V. A Generic Semantic Account of Rule-Following Ascriptions: A Framework for Integrating SSK and ESW

My suggestion is that there is a third way of dealing with the problem of rule-following which may avoid the pitfalls of Bloor and Lynch. The key lies in coming up with a plausible semantic treatment of rule-following ascriptions. I (Cheng, 2011) have proposed that rule-following ascriptions such as "John follows the *plus* rule" be re-characterized as habitual sentences such as "Mary walks to work", which, as I shall explain shortly, are a sub-species of generic sentences like "Birds fly". I further suggest that a non-quantificational semantic analysis of habituals is preferable and plausible, and adopts a cognitive approach modeled after Leslie's treatment of generics (2008). The result of this work is a variant version of the skeptical solution — replacing the truth-condition semantics with an assertibility-condition semantic account — without accepting the skeptical paradox and hence without inheriting its internal difficulties.

My starting point is to recognize that the metaphysical issue of finding a meaning-constituting fact is highly controversial and that whether and how the skeptical paradox can be handled by a straight solution remains indecisive. On this basis, I recommend that we bracket the skeptical conclusion and find an alternative way of understanding what we mean when making rule-following ascriptions. A good place to start is to look at generic sentences which are commonplace in our ordinary language. A generic sentence such as "Birds fly" or "Sharks attack" has the form "As are Bs". It expresses a generalization, which nonetheless tolerates exceptions. Not all sharks attack (actually only a relatively small percentage of them do), and not all birds fly, but we are not prevented from construing "Birds fly" and

"Sharks attack" as true. There is another form of generics, i.e., habitual sentences, which express exception-tolerating generalizations over events. For example, "Mary walks to work" reports Mary's pattern of getting to work. Mary does not always walk to work; she might take a taxi or bus to work occasionally. But deviations from the pattern of walking to work do not block us from construing "Mary walks to work" as perfectly legitimate and true.

I propose that rule-following ascriptions be construed as a form of habituals. The sentence "John follows the *plus* rule" reports a generalization over an exception-tolerating pattern of John's arithmetic behavior. The challenge is, then, to provide a semantic account of habitually construed rule-following ascriptions. Leslie's cognitive approach to generics provides a good model. Seeing that a purely quantificational account encounters enormous difficulties in capturing the semantic behavior of the quantifiers implicitly attached to the underlying semantic structure of troublesome generic sentences such as "Turtles are long-lived" (only a very small percentage of turtles actually live long, but the generic sentence is still construed as true), Leslie (2008) inserts cognitive factors of speakers into the semantic analysis. She proposes that very primitive inductive generalization capacities of human beings be invoked in the processing and production of such generic sentences. This explains why "Turtles are long-lived" is construed as true: a positive instance or two of a turtle's living a long-life would instantly prompt a speaker to assent to the assertions of it.

Leslie's cognitive account of generics is a non-truth-conditional semantic analysis. I believe that the basic idea of it can be suitably employed and exploited to account for habitually construed rule-following ascriptions. I have tried to offer a core set of background conditions under which such an ascription can be made. These conditions involve biological and cognitive factors of a speaker, as well as rich and complex social settings in which the speaker is situated. For example, some horses in circus were trained to count numbers up to a relatively long series, but we are highly reluctant to make the ascription "The horse is counting", simply because horses are not regarded as belonging to the kind of species equipped with suitable brains to count. Background assumptions about the history of schooling, training, life-experiences, etc. also affect how we make ascriptions of rule-following. Our own natively endowed cognitive mechanisms such as having a basic numerical sense described by cognitive neuro-scientists like Dehaene (1997) and Dehaene *et al.* (1999) also significantly influence the ways in which we make rule-following ascriptions. A person who systematically fails to operate single-digit calculations such as $1 + 1 = 2$ or $2 - 1 = 1$ would be instantly excluded as adders.

My cognitive generic account of rule-following ascriptions is a form of assertability-condition semantics, for it describes the conditions under which such ascriptions are made. This account is thus similar to KW's skeptical solution, but does not take the same route. For a rule-following ascription is first and primarily construed habitually and then given a generic semantic treatment, without accepting the skeptical conclusion as KW does. Consequently, this semantic approach does not commit itself to any ontological views of rule-following. In this way, we are allowed to talk meaningfully of rule-following and to make legitimate ascriptions about such phenomena, while leaving intact vigorous work in metaphysics to decide whether rule-following is constituted by community-wide dispositions, individualistic dispositions, or simply an illusion. My claim is that this semantic account of rule-following not only preserves some of the insights of KW and avoids some of its drawbacks, but also represent a position that may well be true to the spirit of Wittgenstein's own rule-following discussions. As Wittgenstein maintains, if someone wants to advance a metaphysical thesis, no one can dispute with him. Wittgenstein has distaste for engaging in metaphysical inquiries. A cognitive generic semantic account of rule-following ascriptions obviously accommodates this negative attitude.

What implications does this semantic treatment have for SSK and ESW? As far as I can see, it offers a framework in which both methodologies are comparable. This is because a cognitive generic semantic account of rule-following ascriptions is neutral about the ontology of rule-following. Therefore, it leaves untouched the issue of whether a community-wide dispositional theory on which Bloor's SSK is based or a non-scepticist view of rule-following on which Lynch's ESW relies is more correct or compelling. This semantic framework gives ample room for either account of rule-following to revise and improvise. As a result, both SSK and ESW can be accommodated in it.

Since a cognitive generic account involves depictions of a huge set of background conditions ranging from biological, cognitive to cultural and social factors of speakers, an impression is inevitably imposed on us that the job of depictions looks very much like giving some causal explanations. This feature may help explaining why both Bloor and Lynch agree, either somewhat impatiently or uneasily, that their methodologies and indebtedness of Wittgenstein do not differ much in many aspects. As Lynch says:

> David Bloor wonders why I am so preoccupied with the differences between ethnomethodology and the sociology of scientific knowledge, and I imagine many readers also wonder. The two approaches certainly have much in common. As far as most sociologists are concerned, the very fact that Bloor and I can argue about the sociological implications of Wittgenstein's philosophy testifies to our sharing an arcane interest (1992, p. 283).

If my proposal has any plausibility, it appears to deliver a position which Wittgenstein may well endorse. Namely, both SSK and ESW are permissible methodologies in the sociology of science. After all, they may emphasize somewhere different aspects of what get involved in scientific practices, and why not simply allow both methodologies to supplement each other. This approach would not only be a safe way to do sociology of sciences, but a productive way.

Acknowledgement

Aside from the conference in Shanxi University, earlier visions of this paper were presented in the Annual Meeting of Korean Society for the Philosophy of Science, and in the Conference on The semi-centennial of Kuhn's Structure: From HPS to STS at Yang-Ming University in Taiwan. I am indebted to helpful comments from the audiences, especially from Rueylin Chen, Inrae Cho, Dawei Fu, Joongol Kim and Michael Strevens. The completion of this paper is made possible by the funding of a three-year NSC research project of Taiwan (NSC 99-2411-H-194-012-MY3).

References

Baker and Hacker (1984). *Wittgenstein, Rules, Grammar, and Necessity.* Oxford: Blackwell.

Blackburn, S. (1984). The individual strikes back. *Syntheses*, 58, 281–301.

Bloor, D. (1973). Wittgenstein and Mannheim on the Sociology of Mathematics. *Studies in the History and Philosophy of Science* 4: 173–91.

______ (1976). *Knowledge and Social Image.* London: Routledge and Kegan Paul.

______ (1983). *Wittgenstein: A Social Theory of Knowledge.* New York: Columbia University Press.

______ (2004): "Institutions and Rule-Scepticism: A Reply to Martin Kusch", *Social Studies of Science* 34/4: 593–601.

Boghossian, P. (1989). The rule-following considerations. *Mind*, 98, 507–49.

Cheng, K. Y. (2011). "A New Look at the Problem of Rule-Following", *Philosophical Studies* 155: 1–21.

Dehaene, S. (1997). *The number sense: How the mind creates mathematics.* New York: Oxford University Press.

Dehaene. S., Spelke, E., Stanescu, R., Pinel, P., & Tsivkin, S. (1999). Sources of mathematical thinking: Behavioral and brain-imaging evidence. *Science*, 284, 970–974.

Garfinkel, H. (1967). *Studies in Ethnomethodology.* Englewood Cliffs, N.J.: Prentice-Hall.

______ (1986). *Ethnomethodological Studies of Work.* London: Routledge and Kegan Paul.

Garfinkel, H. and Sacks, H. (1970). On Formal Structures of Practical Actions. pp. 337–66 in J. C. McKinney and E. A. Tiryakian, eds., *Theoretical Sociology: Perspectives and Development.* New York: Appleton-Century-Crofts.

Hoffman, P. (1985). Kripke on private language. *Philosophical Studies*, 47.

Kusch, M. (2004): "Rule-Scepticism and the Sociology of Scientific Knowledge: The Bloor-Lynch Debate Revisited", *Social Studies of Science* 34/4: 571–91.

______ (2004). "Reply to My Critics", *Social Studies of Science* 34/4: 615–620.

______ (2006). *A sceptical guide to meaning and rules: Defending Kripke's Wittgenstein.* Stocksfield, UK: Acumen Publishing.

Leslie, S. (2008). Generics: Cognition and acquisition. *The Philosophical Review*, 117, 1–47.

McDowell, J. (1984). Wittgenstein on following a rule. *Synthese*, 58(3), 325–63.

McDowell, J. (1992). Meaning and intention in Wittgenstein's later philosophy. *Midwest Studies in Philosophy*, 17, 40–52.

McGinn, C. (1984). *Wittgenstein on meaning.* Oxford: Basil Blackwell.

Shanker, S. G. (1987). *Wittgenstein and the Turning-Point in the Philosophy of Mathematics. Albany: SUNY Press.*

Wittgenstein, L. (1953). *Philosophical investigations.* Oxford: Basil Blackwell.

______ (1967). *Remarks on the Foundations of Mathematics.* Oxford: Blackwell.

MODEL ERROR AND ENSEMBLE FORECASTING: A CAUTIONARY TALE

SEAMUS BRADLEY, ROMAN FRIGG, HAILIANG DU
and LEONARD A. SMITH

1. Introduction

Many scientific enterprises nowadays involve using computer simulations to model complex phenomena of interest. Many models make probabilistic predictions using the methodology of *initial condition ensemble forecasting*, sometimes called ICE forecasting. Weather models, climate models, financial market models and hydrological models, among others, all do this sort of thing. We will see (using a simple example) how this a methodology successfully deals with certain kinds of uncertainty. The main aim of this paper is negative, however. We will show that ICE forecasting does not help with a certain other kind of errors, namely with structural model error. We argue that if a model is chaotic and if there is only the slightest model imperfection, then treating model outputs as decision relevant probabilistic forecasts can be seriously misleading.

Models of the systems mentioned above are extremely complex. The problem we want to discuss shows up in much simpler models, so the remainder of our discussion will be about a particular simple model-system pair that we begin to introduce in the next section. After that, in Section 3 we introduce ICE forecasting and in Section 4 we demonstrate a problem with it. We relate our discussion of the simple model back to weather and climate models in Section 5 and we respond to some objections in Section 6.

2. Basic Dynamical Systems: the Toy Model

There is a pond in the garden that contains a population of fish. The fish population changes over time, and we would like to be able to predict those changes. So we are going to build a model of the pond. We describe the possible states of the system as a set X which we call the *state space*. This is a set that contains all the states the system (the pond) could be in. In the current example, the state space will be possible levels of fish population. Next we need to describe how the state of the system changes over time. We do this with a function we call ϕ_t which is the *time evolution* of the system. If the system is in state $x \in X$ now, the system will be in state $y = \phi_t(x)$ at a later time t. In other words, ϕ_t tells us how the system's state changes over time. So ϕ_t is a function that, given as input the current state of the system, outputs the state of the system at some future time. These elements, X and ϕ_t, are two of the three ingredients of a *dynamical system*. The third element is a measure of the state space, μ. The measure μ assigns sizes to subsets of X.

We now want to look for a dynamical system that describes the population dynamics of the fish pond. To this end we introduce the population density ρ, the actual number of fish in the pond divided by the pond's maximum carrying capacity. So ρ_t is a number between 0 and 1. We adopt the following model [9]:

$$\rho_{t+1} = 4\rho_t(1 - \rho_t) \tag{1}$$

This equation gives us next week's fish population as a function of this week's population. The more fish there are this week, the more there will be next week (up to a certain point): hence the ρ_t term. As the population approaches one half of its maximum, this slows the growth of the population: hence the $1 - \rho_t$ term.

This equation is the well-known *logistic map*.[1] This is an example of a dynamics that demonstrates the property of *sensitive dependence on initial conditions*.[2] What this means is that fish populations that start off being very close together can diverge radically. That is, even if ρ_0 and ρ'_0 are very close, it can be that ρ_t and ρ'_t are radically different for some later t. This means that such systems appear to become unpredictable, unless you know the *exact* initial conditions ρ_0. The next section discusses an interesting methodology used to overcome this problem.

3. Initial Condition Ensemble Forecasting

Unfortunately we don't know the exact current fish population. We don't know what value of ρ_0 to plug in to get our prediction model going. We do have a rough idea of the current population, but we can't be sure exactly which is the "true" ρ_0. Given that the model exhibits sensitive dependence on initial conditions, just using our "best guess" of the initial condition guarantees no reliable prediction. It has become customary to use an ICE to deal with this problem (see, for instance, [1] and [5]). Let's say we take our best guess of the initial condition, and we centre some sort of "noise distribution" around this. The idea is that this distribution represents our uncertainty about the initial condition. What we want to do is to see how this distribution evolves through time.

---

[1] For more on the logistic map see [9] and [10].

[2] For discussion of the concept of SDIC and its relation to predictability, see [10 Chapters 2, 9] and [11 Chapter 4].

So, instead of subjecting only our initial state x to the dynamics ϕ_t, we take some probability distribution around x, call it $p_0(x)$. Ideally we would subject that distribution to the dynamics. This gives us a new probability distribution $p_t(x) = \phi_t[p_0(x)]$. We shall leave aside issues relating to how one chooses p_0. Let's say that it is chosen somehow to best represent the state of uncertainty with respect to the variables in question. In our simulations we make the standard assumption that the distribution is a $U(0, 10^{-2})$, a uniform distribution with range 10^{-2} around x. But this assumption does no real work: similar results would obtain with other kinds of distributions. Our focus is on p_t. The implicit assumption that many make in working with p_t is that it represents a decision relevant probabilistic forecast of the target system. We call this the *default position.* Note that the default position does not make the trivial claim that p_t satisfies the mathematical axioms of probability. The default position is the stronger claim that the probability that this methodology produces can and should be used as the basis for decision making.

In practice we don't know how to subject the distribution to the dynamics analytically. So what we do is sample from the distribution, and subject each member of the sample to the dynamics. The initial conditions sampled from p_0 are called an ICE. Each element of the ensemble is a possible initial condition for the system. We put every ensemble member through the dynamics and thereby create a collection of outputs. The set of outcomes of subjecting members of the ICE to the dynamics is called an ICE forecast. The distribution of outputs of this process is taken to represent the evolution of the distribution. In what follows, we sample 1024 ensemble members from the above distribution centred on a particular initial condition, and we run each member through our dynamics. This distribution of ensemble members is then interpreted as the probability distribution of the future fish population.[3]

Is this a good methodology? Does this give us the right probability? First, it isn't clear what "the right probability" means: we are talking about a deterministic system, and the choice of noise distribution was somewhat arbitrary. So what does it mean to say that this method spits out the correct probability? Here's one way to assess whether the method is sound. Let's see how quickly agents who bet using their ensemble probabilities go bankrupt. We do the betting as follows. We split the state space up into a number of "bins",[4] and we use $p_t(x)$ to set the odds for the actual population ending up in that bin after t weeks. Then various bettors decide which bets to take. If the house's model probabilities are "good", then the bettors shouldn't be able to make huge sums of money off the house quickly and consistently.

The bettors in this scenario are betting simply according to what odds are being offered. That is, one agent will bet on an event if the inverse of the odds — the

[3]For further discussion of the interpretation of ensembles, see [5] and [4], and the other papers in that thematic issue of the Philosophical Transactions of the Royal Society.

[4]In our simulations there are 32 bins.

"implied-probability" — is in (1/2, 1]. Another bets on an event if its implied-probability is in (1/4, 1/2], and so on.

Simulations show that these bettors don't make consistently huge gains. The house's model-based probabilities have therefore passed this test. So this is a promising methodology. The probabilities generated by the ICE forecast can usefully be taken as the basis for decision making.

4. Model Error

Sadly, it's not all good news. The spanner in the works is structural model error.[5] Let's imagine that the fish population dynamics is not given by the above model and that the actual dynamics are represented by the following equation:

$$\tilde{\rho}_{t+1} = (1-\epsilon)4\tilde{\rho}_t(1-\tilde{\rho}_t) + \epsilon\frac{16}{5}\left[\tilde{\rho}_t(1-2\tilde{\rho}_t^2+\tilde{\rho}_t^3)\right] \tag{2}$$

As Figure 1 shows, the curves described by Equations 1 and 2 are very similar.[6] In fact, the maximum one-step error of the model is 5×10^{-3}, which is very small. One might expect that such a small error means that the model ensemble is still a pretty good guide to how to set your betting odds. Unfortunately, this is not the case.

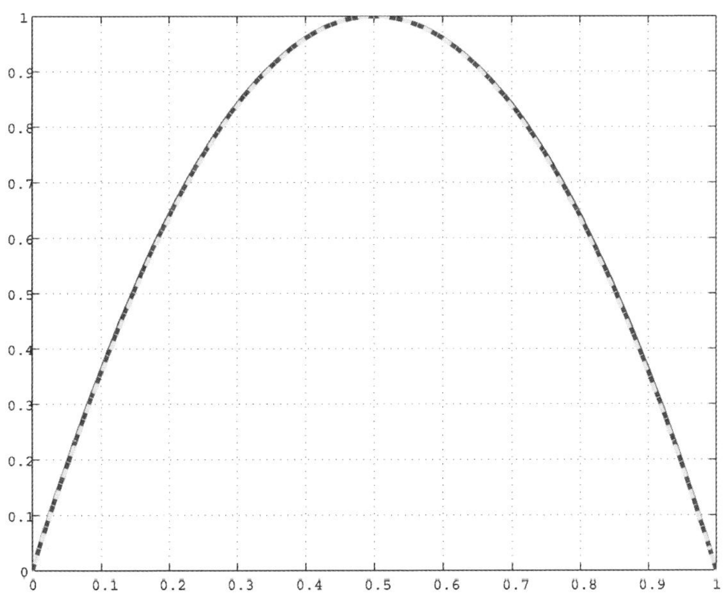

Fig. 1. Graphs of Equation 1 (hatched line) and Equation 2 (drawn line).

[5]We face structural model error when the model's functional form is relevantly different from that of the the true system.

[6]The figures in this paper are reproduced from [7].

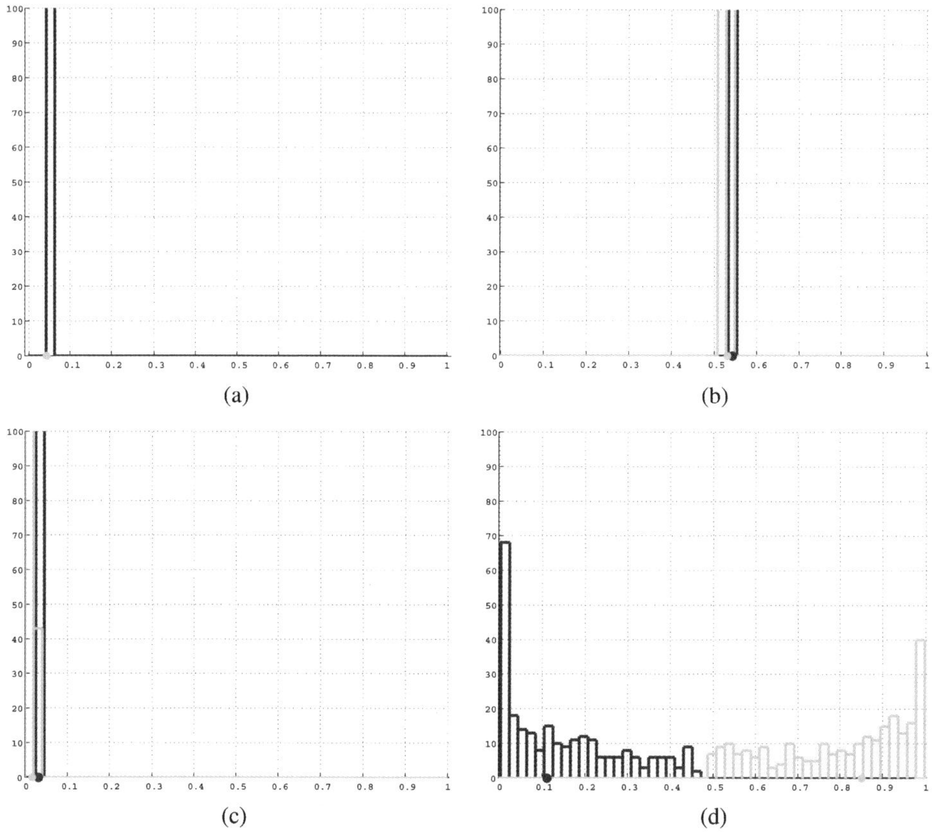

Fig. 2. Histograms of ensemble members at lead times $t = 0, 2, 4, 8$.

We ran 1024 samples from our initial distribution through the dynamics of the system and through the dynamics of the model. We then produced histograms of number of ensemble members in each bin after 1,2,4 and 8 weeks.

As Figure 2 demonstrates, the system distribution and the model distribution can come apart radically in the middle term. In the short term, the distributions overlap, since the model error has not had time to have its effects blow up. In the long term, both distributions settle down into looking roughly similar (they have effectively converged on their invariant measures, which appear to be similar).[7] But in the middle term, the distributions look wildly different!

It might be obvious that using the model-based probabilities as probabilities for decision-support would be unwise in this case, and probably disastrous. Consider the house's odds for the event "$\rho_8 \leq \frac{1}{2}$". That is, the event that the fish population will be less than half its maximum value after eight weeks. Almost all the model

[7]We have not shown figures of the long term, but by about $t = 16$ the distributions come to overlap again.

probability is in the left half of the $t = 8$ histogram. So the house sets short odds on this event, and very long odds on "$\rho_8 > \frac{1}{2}$". As the system probabilities show, however, almost all ensemble members actually land in the right half of the graph after 8 weeks. So it looks as if a house setting its odds on the model probabilities would lose a lot of money: it is putting very long odds on an event that is actually very likely to happen.

Simulations back this point. In the scenario discussed in [6], there are nine bettors. One will bet on an event if the inverse of the odds — the "implied-probability" — is in $(1/2, 1]$, another bets on an event if its implied-probability is in $(1/4, 1/2]$, and so on. They are simply betting according to what odds are being offered. They have no information that the house does not have. In other words, the information is completely symmetrical. The simulations show that the bettors mostly very quickly increase their wealth, thus the house must be losing money. But if bettors who know no more than the house can rapidly increase their wealth, then there must be a serious problem with how odds are being offered.

One might push back against our worries by arguing that we have biased the discussion by a judicious choice of initial distribution. Sadly this is not the case. The calculations were repeated with 2047 different initial distributions. About a third of them behave in a way similar to the one seen in Figure 2; in about another third of the cases the two distribution came unstuck albeit in less dramatic fashion than in Figure 2; and in the last third of the cases the two distributions remained relatively close. So this problem is not one isolated to a particular initial distribution!

The conclusion is that offering odds corresponding precisely to model-based probabilities derived from a model with structural model error can have disastrous consequences and is therefore better avoided. There is a temptation to respond that this does not show that probabilities are useless; it only shows that we should not use these probabilities when they are misleading. The problem with this suggestion is that outside a thought experiment we have no means to tell when that happens. The only thing we have is the model which we know to be imperfect in various ways. We know that model-probabilities and probabilities in the world can come unstuck dramatically, but we don't know where and when. So while there is definitely information in the model-based probabilities it would be foolish to take them as actionable probabilities in decision support. As long as we have no means of of coping with the their shortcoming, we'd better be on guard!

5. Weather and Climate Modelling

Recall that a dynamical system consists in a state space X which describes the possible states of the system, and a time evolution ϕ_t which describes how the system changes in time. In the case of a weather or climate model, the state space X will consist in a description of relevant atmospheric variables (air temperature, wind speed, humidity, precipitation and so on). To fully describe the weather or climate system we would need to specify the values of all these variables at all points of the atmosphere. Such an endeavour is effectively impossible: first, even given the assumption that these quantities are well defined, we don't have enough measuring devices to keep track of every variable at every location in the atmosphere, second we want to use computers to simulate the climate and the storage capacity of these devices is finite. We might therefore discretise the state space by putting a finite grid on it and describe the relevant variables in only a coarse-grained way. Even the smallest grid used in modern global climate models are about 100 km on a side. This is a rather coarse description. For instance, the entire city of London is now represented by one set of numbers (one number for temperature, one for precipitation etc.). On top of this, the dynamics of the model introduces its own distortions. The ϕ_t for a climate model will involve many idealising assumptions. Important aspects of the topography of the surface of the Earth get distorted: the resolution of the models does not allow for realistic mountain ranges like the Andes; the Southern half of the state of Florida may be missing; many islands simply do not exist; clouds cannot be resolved properly.[8] Beyond merely failing to resolve certain processes directly, there is a deep question about what exactly is being represented in the model: there is no such thing as the true wind speed in the model grid point corresponding to central London.

Both weather and climate models thus suffer from various kinds of model error.[9] So the above cautionary tale should suggest that using ICE model-probabilities as decision relevant may be dangerous. And yet, it seems to be assumed by many people that such ensemble probabilities do provide decision-relevant information.

A recent example is a project entitled UKCP (United Kingdom Climate Projections), which aims to make high resolution probabilistic projections of the future climate for the United Kingdom for up to the end of the century.[10] It predicts, for instance, with a probability of 0.5 that there will be a 20–30% reduction in precipitation in London by 2080. These predictions are remarkably precise. The high resolution of the grid on which the predictions are made is able to distinguish the effects of climate change in London and Oxford (which are only an hour apart by train).

Developing and operating these large computational models is a significant cost to the groups who do it, and an increasing number of scientists are devoting their careers to building and running such models. The question naturally arises whether these models can deliver as advertised. The insights gained in the last section would urge both some caution in decision making and some care in resource allocation in the modelling process.

[8]There will of course be parameters that attempt to simulate the effects of things like clouds.
[9]For more on uncertainty in science, see [3].
[10]UKCP's probabilities are derived in a more complicated way than the ones discussed in the last section, but the issues we describe equally apply to more complex schemes. For a discussion of UKCP's methodology see [8].

6. Ways Out?

In this section we discuss various ways one might try to avoid our negative conclusions. First, we return to the question of whether we are interested in the "middle term" — where the distributions come apart drastically. Some have objected to our argument by saying that we are in fact only interested in short term predictions and thus our results do not pose a problem.

Of course we may sometimes be interested in the short term, but in both weather and climate modelling we also are interested in the medium or long term behaviour: we do not limit predictions to short lead times. What counts as short-term or long-term is relative to the model and it could be the case that by standards of the relevant climate models a prediction for 2080 is still a short term prediction. We are doubtful that this is the case. Indeed, it would be surprising if such predictions would turn out to be short term by the lights of a model used to make that prediction, in particular given that state of the art climate models differ even in terms of their performance over the past century. So we take it that the burden of proof lies squarely with those who believe that this is the case.

Going the other way, some critics argue that one only ought to be interested in the long term behaviour of the system. Such behaviour is dominated by the structure of the invariant measure, and so all we need to do is study the invariant measure. Implicit in this proposal is the assumption that the invariant measures of the system and the model are similar. While this assumption may be true for the toy example discussed above, it is not true in general. First, we have no reason to assume that the non-linear systems we are interested in are *structurally stable* in the sense that similar systems have similar invariant measures. Second, the actual system is a transient system (that is, when the climate is changing) and as such does not have an invariant measure at all.

A further possible defence of the default position might just take our results to be inapplicable to actual modelling practices because weather and climate models are very different from our simple logistic map case. We agree that the models are very different, but as we saw in the last section, weather and climate models are likely to have even more serious model errors than in our toy example. We think the burden of proof is on the proponent of the default position to show what it is that makes such models immune from our criticism. The weather and climate modelling methodology is obviously much more complex than the simple case we presented above, but are any of the relevant differences such as to invalidate the applicability of our conclusions?

It is fair to say that there is no hard and fast argument for this conclusion. It seems to us, however, that the burden of proof lies with those who want to argue that the logistic map is a special case that the default position does not run into the problems we describe when used in the context of other non-linear models.

References

[1] Anderson, Jeffrey L. Selection of initial conditions for ensemble forecasts in a simple perfect model framework. *Journal of the Atmospheric Sciences*, 53(1): 22–36, 1996.

[2] Argyris, John, Gunter Faust and Maria Haase. *An exploration of chaos.* Elsevier, 1994.

[3] Bradley, Seamus. Scientific uncertainty: A user's guide. Technical Report 56, Grantham Institute Working Paper, 2011.

[4] Collins, Matt. Ensembles and probabilities: A new era in the prediction of climate change. *Philsophical Transactions of the Royal Society*, 365: 1957–1970, 2007.

[5] Leutbecher, Martin and Tim N. Palmer Ensemble forecasting. *Journal of Computational Physics*, 227: 3515–3539, 2009.

[6] Frigg, Roman, Seamus Bradley, Hailiang Du and Leonard A. Smith. Laplace's demon and the adventures of his apprentices, forthcoming in Philosopy of Science (January 2014).

[7] Frigg, Roman, Seamus Bradley, Reason L. Machete and Leonard A. Smith. Probabilistic forecasting: Why model Imperfection is a poison pill forthcoming, in Hanne Anderson, Dennis Dieks, Gregory Wheeler, Wenceslao Gonzalez and Thomas Uebel (eds) *New Challenges to Philosophy of Science.* Berlin and New York: Springer.

[8] Frigg, Roman, David Stainforth and Leonard A. Smith. The myopia of imperfect climate models: the case of UKCP09. *Philosophy of Science*, 2013.

[9] May, Robert, Simple mathematical models with very complex dynamics. *Nature*, 261: 459–467, 1976.

[10] Smith, Leonard A. *Chaos: A very short introduction.* Oxford University Press, 2007.

[11] Smith, Peter. *Explaining chaos.* Cambridge University Press, 1998.

THE CONTEXTUALIST BASIS OF SCIENTIFIC EXPLANATION

GUICHUN GUO
Research Center for Philosophy of Science and Technology
Shanxi University

From perspectives of contextual structure, elements and features, this paper analyses the contextualist basis of scientific explanation. It is articulated that contextual analysis specifies the boundary conditions of syntactics, semantics, and pragmatics, clarifies the essence and value-orientation, constructs a conversational platform, and unveils the progress condition and historical continuity of scientific explanation.

As a central topic in study on contemporary philosophy of science, scientific explanation attracts almost all the philosopher's interests and discussions in various levels and perspectives for more than sixty years. Until now, in the first decade of the 21th century, this issue remains a core focus around the international academic community of philosophy of science. The published papers relevant with scientific explanation nearly account for half the number in professional journals which are mainstream ones in the discipline, such as *Philosophy of Science*, *The British Journal of Philosophy of Science*, and *The Journal of General Philosophy of Science*. But there also is an interesting and thought-provoking phenomenon: the academia doesn't have a unanimous agreement with each other on the nature of scientific explanation. Based on their own specific characteristics, different explanation models provide particular insights on the issue of scientific explanation from different perspectives. In such a manner, a phenomenal situation emerges: different models are in the condition of competition and opposition, they isolate each other and therefore basically in unending quarrel and fighting. There has not yet been an effective approach to provide them an appropriate platform for communication and integration. Through a kind of contextualist analysis on the structure, elements and features of scientific explanation, this paper elaborates its essential context-dependence and context-sensitivity. Thus, on the contextualist basis, a possible way to construct a uniform theory of scientific explanation is explored.

I. The Contextual Structure of Scientific Explanation

The traditional theories of scientific explanation aim to establish a kind of formalized model, especially trying to do so by means of syntactical and semantic analysis. Virtually, suchlike models almost all presuppose a stable and fixed contextual structure of scientific explanation. That is, scientific explanation is the description and reproduction of conventionalized entity of scientific knowledge. From 1970s onwards, philosophers of science such as Van Fraassen more and more clearly realize that this kind of understanding of contextual structure is formalized one-dimensionally, and thus is not complete and sufficient. Only when having taken the long-ignored pragmatical analysis into consideration, it is possible for us to describe integrally and present completely the contextual structure of scientific explanation, then to understand and grasp comprehensively its essence.

1. The Syntactical Basis of Scientific Explanation. The syntactical basis of scientific explanation is the logical and syntactical form embodied in the language of scientific explanation. The classical paper, *Studies on the logic of Explanations*, by Carl G. Hempel and Paul Oppenheim intended to explore this primary aspect of scientific explanation. In this paper, scientific explanation was dealt with through rigid formal analysis, and the distinguished Deductive-Nomological model is presented. It opened the prelude of the contemporary studies on this key issue in philosophy of science. After this breakthrough, many eminent philosophers of science, including E. Nagel and K. Popper, began to publish relevant literatures in succession to express their approval and support. Providing a very important foundation for the traditional understanding of scientific explanation, the D-N model immediately became a prominent paradigm as soon as it was proposed. Based on the classic first-order logic and under adoption of pure method of syntactical analysis, the D-N model made the following assumption to the syntax of model language L in scientific explanation: "L has the syntactical structure of the lower functional calculus without identity sign. In addition to the signs of alternation (disjunction), conjunction, and implication (conditional), and the symbols of universal and existential quantification with respect to individual variables, the vocabulary of L contains individual constants ('a', 'b', $\cdots$), individual variables ('x', 'y', $\cdots$), and predicates of any desired finite degree; the latter may include, in particular, predicates of degree 1 ('P', 'Q', $\cdots$), which express properties of individuals, and predicates of degree 2 ('R', 'S', $\cdots$), which express dyadic relations among individuals. For simplicity, we assume that all predicates are primitive, i.e., undefined in L, or else that before the criteria subsequently to be developed are applied to a sentence, all defined predicates which it contains are eliminated in favor of primitives. The syntactical rules for the formation of sentences and for logical inference in L are those of the lower functional calculus. No sentence may contain free variables, so that generality is always expressed by universal quantification." (Hempel & Oppenheim) Emphasizing that scientific explanation essentially is syntactical, this opinion expresses a typical point of view of logical positivists on scientific explanation. The purpose of Hempel and Oppenheim is to abstract the pure logical and syntactical elements from the concept of scientific explanation, so as to peel off the pragmatic factor. Thus the relativity problems about the individuals concerned in the explanation process are avoided. The D-N model sets the most fundamental syntactical boundary for scientific explanation. That is to say, it is impossible that specific context of scientific explanation can surpass the syntactical boundary of a given language. Especially for the formalized objects in, for example, mathematics and physics, their contexts must be restricted by the syntactical periphery of relevant logical syntax and formal algorithm. In this sense, the grammatical extent of scientific explanation marks syntactical borders of explanatory context. More complete is the axiomatization degree of scientific theory, more clear is the syntactical border of its explanatory context.

2. The Semantic Rule of Scientific Explanation. Even though the syntactical boundary has been specified as a primary task of *Studies on the logic of Explanations*, the contextual structure still needs to be clarified on the level of semantics. First, some central concepts such as "theory", "*explanandum*", "*explanans*", "law", "law-like sentence" should be described semantically, so as to get the relevant semantic stipulation. Second, the semantic rule of model language L should be fitted into corresponding assumption. Hempel and Oppenheim made a basic description of the semantic rule for scientific explanation as the following: "The primitive predicates of L are all purely qualitative. The universe of discourse of L, i.e., the domain of objects covered by the quantifiers, consists of all physical objects, or of all spatio-temporal locations." (Hempel & Oppenheim) This rule defines the semantic borders of certain representations for the systematic value trends within relevant explanatory context. It is that the semantically constructive principle stipulates the extent of the semantic tension, establishes the elasticity degree and intention value of semantic explanation in certain context. Furthermore, it is that the semantically constructive principle enforces the inner relevance between the items, propositions represented by specific theories and objects, and the various respects of the world referred by them. The semantic rule combines with the syntactical basis, the two jointly determine the objective side of scientific explanation, emphasize that scientific explanation should make claims having common sense in a formal mode. The semantic rule of scientific explanation gives proper representation for the *explanandum*, determines that the generality of scientific explanation can be discussed critically and tested ultimately by different scientific communities or individuals. That is the source of the "explanatory power" of scientific explanation. As core in the objective contextual structure of scientific explanation, the semantic rule marks the precondition of the emergence of scientific explanation in the intersection of objective and subjective contexts. The semantic rule endows the practical scientific explanation with meaningful stipulation, and the preconditions and regularities of explanation depending on the semantic level of propositional explanation. Hence, based on the syntactical basis, the semantic rule constitutes the most important and central part of premises set of the context in scientific explanation.

3. The Pragmatic Boundary of Scientific Explanation. Manifesting in certain process and procedure, scientific explanation primarily presents itself as a kind of scientific practice. To some extent, it is realized through the form of linguistic acts. According to Peter Achinstein, explanation may refer either to "a process" or "a product": "The process is a linguistic performance; someone explains something to someone by uttering or writing statements. The product is the content of the linguistic performance. The linguistic performance itself involves an intention on the part of person producing the explanation. " (Salmon, 146) Since the same set of statements or linguistic sequences can be used "either to explain or do other sorts of things", when trying to define and evaluate the explanatory product, we have to bring the contextual elements in consideration, especially the explanatory intentionality and the illocutionary force. For instance, in the circumstance where the intentionality installs different pragmatic boundary, the linguistic sequences or argument may function as explanation in certain context, but as prediction in another. It is the determinacy of pragmatic boundary rather than any single command of generality makes judgment and evaluation to the advantages and disadvantages of scientific explanation. The expansion and contraction of pragmatic boundary will lead to the changes of explanatory power. More specifically, a given scientific explanation A explains certain fact X to certain object P_1. Meanwhile, for another object P_2, it is possible that the same explanation can not make a proper explanation about X. P_2 even possibly does not think that X need to be an explanation. Or P_2 may find out that X is not possible to be understood. Or A does not have explanatory power. Or A is not relevant at all with X that bothers him. So, in essence scientific explanation expresses a kind of boundary of pragmatical relation. That is what Hempel points out, something which can be said meaningfully just forms a kind of explanation to this or that individual in this sense. (Hempel, 425–426) This profoundly indicates the meaning that pragmatic boundary formulates the sphere of context application in scientific explanation.

In short, the syntactical basis provides the formal ground for scientific explanation, the semantic rule formulates the meaningful frame, the pragmatic boundary sets the scope of application and standard of evaluation. In their entirety, the unification of the syntactical, semantic and pragmatic levels presents the contextual structure of scientific explanation. It should be the primary level on which we understand the contextualist basis of scientific explanation.

II. The Contextual Elements of Scientific Explanation

Based on the clarification of the contextual structure of scientific explanation, it is necessary to elaborate the contextual elements functioning on this ground. That is to say, the syntactical, semantic and pragmatic structures integrate to build the formal base, meaning frame, application scope and evaluation criterion of scientific explanation. When these are determined, what sort of elements of scientific practices can influence the processing of specific details of scientific explanation? How these elements constitute the contextual particularity of scientific explanation as major elements? We argue that these contextual elements mainly consist of the following:

1. The Question Element of Scientific Explanation. The scientific practice is originally guided by scientific question. Therefore, scientific explanation inevitably starts from certain questions. The context of question is the position of actual difficulty or situation of real dilemma in specific phase of scientific research, with which scientists have to confront and deal. At that time, there will be some blind spots of existed knowledge are realized. The context of question proposes a limitation to the hypothesis-formation and experimentation-possibility of a sort of satisfied explanation. That is just as what John Dewey had pointed out: "Demanding for the solution of a perplexity is the steadying and guiding factor in the entire process of reflection. Where there is no question of a problem to be solved or a difficulty to be surmounted, the course of suggestions flows on at random······But a question to be answered, an ambiguity to be resolved, sets up an end and holds the current of ideas to a definite channel. Every suggested conclusion is tested by its reference to this regulating end, by its pertinence to the problem in hand. This need of straightening out a perplexity also controls the kind of inquiry undertaken." (Dewey, 11) Under the guidance and setting of context of question, it is the subjects that develop the explanation process. The result is the production of some ultimately satisfied explanations which can successfully give an answer to the original question. The relevant elements of the context of question include: the presupposition of the required question, the special time-point of the question-formation, the particular form of question adjustment and modification in the elaboration of scientific explanation. First of all, the question which needs to be explained is proposed. Second, the question is transformed into a kind of form of conceptual model, and then forms a set of related propositions. This sort of operationalization of conceptual model necessarily is statistically tested in the limitation of the context of question. Consequently, the question which needs to be explained and its forming process are contained in special conceptual model of the context of question. As Van Fraassen points out, scientific explanation is in essence the answer to a question which is to be inquired. In other words, scientific explanation boils down into seeking the answer to a "why question". Obviously, the why question can only be confirmed in a certain context. At this point, the context of question is determined by a set of background knowledge K. (Kitcher & Salmon, 414) As a sort of answer, scientific explanation needs to be evaluated by the original question. However, the answer lies under the limitation of relevant requirement of contextual information, that is the question "why the situation is P?" As soon as the context changes, the same story goes with the information elements required. Besides that, the data required in question evaluation, the parts in background information which is used to evaluate the answer, both depend on the context of question. (Van Fraassen, 156)

2. The Background Element of Scientific Explanation. The background element of scientific explanation is the theoretical premise based on which explanatory question is proposed. It consists of the background belief of certain research dealt with by the scientific community. In fact, the differences in standards of scientific explanation are different levels of generality of knowledge background, theoretical system, and belief tendency. Scientific explanation requires highly confirmed knowledge background, theoretical paradigm and cognitive conditions. That means, scientific explanation is relative to our belief of whole science rather than some specifically experimental data or statement of evidence. For the intentionality, pertinence and connotatively inferential structure of scientific explanation, the background element has a prominently restrictive function and influence. Scientific explanation is under a multi-level and comprehensive influence of the background elements of scientific knowledge. Among them there are the observation data, the empirical generalization, the certain theoretical hypothesis, and the general picture of the world which is popular in specific stage of science. Some philosophers of science thinks that, whether a scientific explanation has real explanatory power, a necessary condition is whether it follows some kind of special picture of the world. For example, in the background of 19th century physics, if scientific explanation could have some explanatory power, it had to be proceeding on background of the mechanist and corpuscular theory. (Kantorovich) The background element covers the existed knowledge which is to be amended in scientific explanation, and the cognitive context which makes some knowledge amendment becoming explanatory act. Scientists with various theoretical hypothesis and intentionality inclination may bring the *explanandum* into different theoretical frameworks. At that time, the *explanandum* actually is brought under different contexts of background. In a certain background context, the coherence and the explanatory power of the same explanation are increasing, while in another context the relevant properties are weakening. Thus it can be seen that, when determining the united property of some newly presented alternatives of scientific explanation, the structure of contextual background plays a key role. So it should be regarded as the decisive judgment influencing explanatory advantage. According to Peter Lipton, for a given criterion, how distinguished is an explanation partly determined by how is the other explanation accepted, while the criterion itself is determined by the background. "The importance of the background in inference, and the plausible suggestion that what accounts as a lovely explanation is thus context sensitive, is entirely compatible with inference to the Best Explanation..." (Lipton, 139–140)

3. The Subject Element of Scientific Explanation. The subject of scientific explanation includes all the participators involving in it, such as its presenters and receivers. Focusing on a certain question, the subjects of scientific explanation operate the explanatory act under specific circumstance. This act presents itself as a sort of linguistically communicative act between explainer and audience. It has a meaning of double contextualities: on one side, the explanatory act of subject is modeled by the context of question; on the other side, it constitutes the basis for fixation and renewal of the context of question. So, scientific explanation essentially contains the structure of knowledge-information exchange among its subjects. In the linguistic interaction among the participators, the puzzlement of the audience about certain question is dissolved. Of course, the pragmatic analysis and the settlement of pragmatical boundary do not need appeal to relevant subject standard, and can not be boiled down to a relativist explanation theory. (Caruana, 68) First, single explanation, either deductive or statistical one, could not separate itself from the knowledge situation or cognitive status of explaining subject, and thus to be evaluated. The question of scientific explanation appears as a choice by scientists as an individual observing unit. Various subject of explanation may give different meaning to content and indication to the same empirical observation. Based on their context-sensitivity, the subjects of explanation determine what sort of causal information is important one. In other words, for scientific explanation, that what kind of casual relation is important relies on the subject element in explanatory context. It is the subject of explanation determines the role and function of certain explanation in scientific research. Second, scientific explanation necessarily involves the audience, because the explaining procedure is to explain something concisely and clearly to the audience. The result of explanation should be the understanding of the explained question by the audience. Accordingly, only for the competent audience, scientific explanation is possibly to be understood and get its meaning. In other words, scientific explanation has to be accepted wholly, or at least partly, by the audience. And it is also significant that, by means of reflecting the occurring structure of the phenomenon explained, scientific explanation has actual effect on the mind of explanation participators.

4. The Intentionality Element of Scientific Explanation. The intentionality element of scientific explanation is intrinsically related with the subject element, contains its value orientation. The intentionality has a quality of procedure and continuity. On the one hand, intentionality develops and expands the original context of question under the context of background; on the other hand, it dissolves the context of question in its own intentional structure. Hence, guided by intentionality, the succeeding steps of explanation process all linked with one another. While the previous one changing, every step continually obtains some renovation. Scientific explanation is the respond of explainer with certain intentionality to certain question. This intentionality is blended with the intentionality of the other participators. The latter could provide some empirical data and relevant information which otherwise might be ignored. It is the fusion of intentional horizons makes the original question which is not yet understood understandable, or make it which is abstract and complicated concrete and distinct. The intentionality of audience in explanation usually brings about unexpected insight or new perspective for the explainer. The different levels of explaining intention deliver different types of generality, while the different sets of comparative system and types of generality are meaningful just for different intentions of explanation. (Potochnik) The subject intentionality of explanation leads the conclusion of scientific explanation to diverse causal relations. In philosophy of language, as a sort of illocutionary act, explanation is similar with the speech act, such as dissuasion, warning and promising, typically is accomplishes through speaking words with certain intention and appropriate purpose under certain circumstance. In this sense, functioning as a sort of intentional act in construction and communication of scientific theories, scientific explanation presents itself as a special kind of practice of scientific rhetoric.

Scientific explanation origins from the actual scientific question, the proposal and answer of the question could not be isolated with certain theoretical and knowledgeable background. With the unfolding process of scientific explanation, the elements of both question and background combine with the elements of subject and its intentionality. These elements enhance the expanding, integration and renovation of the original contextual structure by the supplement of new contents. The meaning and significance are produced and accomplished on the basis of the connectiveness of special explanatory criterion and its contextual structure. Among them, while the elements of question and background mainly concern with the form of syntactical basis, and the acceptability of semantic rule, the elements of subject and its intentionality mainly concern with the condition and scope of pragmatical boundary in scientific explanation.

III. The Contextual Features of Scientific Explanation

The structure of scientific explanation is a contextualized one. Its constituent elements are context-sensitive and context-dependent. Consequently, scientific explanation should be understood in essence as a contextualized one. The contextual feature of scientific explanation could be revealed as the structure of dynamic contexts, via the combined functions of their constituent elements. Specifically, the contextual features of scientific explanation mainly include the following aspects:

1. The Feature of Contextual Open-endedness. In his *The Context of Explanation*, R. S. Cohen points out that, a real explanation always occurs in a dynamic context. (Cohen, x) When analyzing the contextual features of scientific explanation, first of all we should realize that, an actual question of scientific explanation will never be totally open-ended. Instead, it will be always limited in some extent. This limitation expresses the flexibility of the context of question, and the fluctuation between general description and formal representation of explanatory language. That is to say, scientific explanation is brought about through constant and dynamic process of re-contextualization. This is sufficiently reflected in the point of view which regards scientific explanation as "metaphorical re-description". This opinion was proposed by Mary Hesse. She argues that, at the start point, scientific data are described with the observable language or the similar. In scientific explanation, these data are re-described in metaphorical way, specifically by means of a new model of interaction theory. During this process, various metaphorical perspectives adopted may lead us conduct multi-dimensional insights to the same question or object. At the same time, the language of observation and literal meanings of words originally belonged to the primary system convert to the metaphorical one. Thus, this sort of new meaning is shaped in the conceptual network of certain context. It is just this sort of intersection and interaction among different parts of conceptual network generate newly changed context of explanation. In this new context, we can discover and express the deep analogy among different phenomenon. Meanwhile, there is rarely any kind of strictly deductive relation between the *explanandum* and *explanans* in scientific explanation. Usually, there is a relation as an approximate correspondence. The language of observation and all natural languages are expanding through this sort of metaphorical use, producing the items of the *explanans*. Therefore, the traditional deductive model of scientific explanation is not complete. It should be modified and complemented by a sort of idea that regards theoretical explanation as metaphorical re-description of target domain. The reason why regarding the essence of scientific explanation as metaphorical re-description, and this kind of process of metaphorical re-description is rational and reasonable, is because that scientific rationality precisely contains an important connotation: the language of scientific theory constantly is adjusted so that accommodating the expanding world. (Hesse, 123) Second, we should notice that there is a very intimate relation between scientific explanation and scientific understanding. In some extent, the purpose of scientific explanation is to express or deliver a sort of special understanding of the world. In a given context of scientific explanation, the participators all have a specific domain of understanding. The task of explanation is the proper fusion of horizons among different domains of participators' understanding. Seeking a sort of scientific explanation presupposes the understanding of something, is a connection between research objects and understanding domain through a proper way. In fact, whatever the difference there is, all kinds of ideas on scientific explanation think that explanation is closely related to understanding. The most prominent practical characteristics of scientific explanation and the understanding it leads to is the guidance to action. We use the result of understanding as a kind of symbol, stopping to seeking further explanation, and then proceeding action according to this kind of explanation belief. (Trout) Hence, one can see that, the basic logic of scientific explanation is not deductive. It is the understandability, based on exploration and history, constitutes the essential characteristic of the logic for scientific explanation. All the various contexts of scientific explanation necessarily contain the elements of understanding, and every element of understandability in special context has its own characteristic. In other words, every context of scientific explanation has its own characteristic form of understanding.

2. The Feature of Contextual Systematicness. First, as stated above, scientific explanation could not be simply limited in the syntactical consideration which only concerns relevant elements of formal model. It is necessarily that the semantic intimacy, rationality and coherency are introduced, and it also depends on the pragmatic consideration which concerns relevant elements, such as explanatory subjects and their intentionality. At the same time, scientific explanation contains the social form of explained event, even the imaginary and inspirational contents of the explanatory idea. Thus, we argue that the essence of scientific explanation is contextualized: the contextual elements such as background, question, subject and its intentionality jointly constitute a specific contextual structure in dynamic way. That is to say, the general mode of scientific explanation or the common form of explanatory model should be a sort of systematic contextuality. Second, the development of science is drove significantly by the desire on scientific explanation. Scientific explanation brings the scientific phenomenon into a unified picture of science which we could be able to access. When satisfied a need of specific scientific explanation, the substantial progress of scientific theory is achieved. That is a very important characteristic which makes scientific explanation having cognitive meaning: scientific explanation systematically leads to the continuous production and development of scientific theory. In the theoretical construction, scientific explanation which could effectively retain, promote and put forward the progress of theories, is regarded and maintained as good explanation. The primary reason why the good or successful explanation could have strongly explanatory power and conspicuously satisfied effect, is that it is constituted by a set of explanations, the alleged "explanatory store" by Philip Kitcher. According to Kitcher, the basic task of explanatory theory is to clarify the contextual boundary and specify the conditions on the explanatory store. "...the explanatory store associated with science at a particular time contains those derivations which collectively provide the best systematization of our beliefs. Science supplies us with explanations whose worth cannot be appreciated by considering them one-by-one but only by seeing how they form part of a systematic picture of the order of nature." (Kitcher & Salmon, 430) Third, under the guidance of specific view of the world, scientific explanation has to make reasonable conjecture and hypothesis to the composition and structure of the world. That is from the explanation of involved ontological frame and cognitive level to the construction of specific theory, and finally reveals the fundamental manifestation of the context-dependence. Primarily, scientific explanation needs to choose a sort of specific theory, and then seek a proper model within this theory. Only under this condition, it is possible that scientific explanation derives the existing form and processing mode of phenomenon in various levels, for instance the ontological and cognitive levels, from the primary principle. While every theory and every model have their own limitation of validity and element of idealization, in order to make the explainer sure about its reliability, they have to be brought into the consideration on the explanatory store. Besides, not all the characteristics of explanation have to be brought into serious discussion. The contextual elements such as question, background, subject and its intentionality will comprehensively classify and filter these characteristics: "Selection of introduced features for explanation is a different and pragmatic, context-dependent matter; we often indicate such selection by stress, emphasis, and so on." (Ruben, 169)

3. The Feature of Contextual Cross-sectionness. To begin with, the theory of scientific explanation develops as a historical evolution from the D-N model, deduction-statistical model to the induction-statistical model, from the historical explanation to the teleological explanation, from the functional explanation to the explanatory pragmatics. The current studies on scientific explanation move forward to new issues such as mixed-level explanation and pedigree explanation. It embodies an important turn: from logical to pragmatic, from rational to explanatory diversity, from explanation to interpretation and prediction. In this process, it is the contextual analysis that could mark the syntactical, semantic and pragmatic borderlines of scientific explanation, so determine the conditions which the meaning of scientific theory relies on, and clarify the value orientation of scientific explanation, reveal the meaning of scientific theory. Furthermore, the contextual analysis describes the advancement conditions and historical continuity of scientific explanation, makes its interpretation of possible worlds possible. It also constructs a platform, letting different schools and ideas communicate each other and merge together, vividly showing the feature of contextual cross-sectionness of scientific explanation. Next, the feature of contextual cross-sectionness of scientific explanation is prominently reflected in the following changes of theories of scientific explanation: from structure to process, from part to the whole, from experimental science to cognitive science, from construction to network model, from truth to the maximum description of objectivity of ultimate research. The most notable characteristic of this turn more clearly highlights the change of frames from object to structural relationship, since the feature of contextual cross-sectionness is distinguished in grasping meaning of scientific explanation. In fact, the conceptual frame or representational mode is achieved through this kind of contextual cross-sectionness. Finally, the contextual cross-sectionness of scientific explanation is embodied in its continually permeating, stretching and expanding from natural sciences to social sciences. A kind of unified theory of scientific explanation should be having very robust comprehensiveness, explaining both natural sciences and social sciences. Although there is remarkable difference between natural science and social science, it does not mean that they necessarily should be having entirely different explaining models. For instance, there are some common contextual features between biological explanation and sociological explanation. In this case, the traditional gap between natural and social science disappears. Therefore, based on contextualism, the idea of scientific explanation provides an important bridge for the methodological fusion between natural and social science studies. (Gonzalez)

As a scientifically cognitive process, scientific explanation makes amendments to the original knowledge in two levels. For one thing, as a sort of complete system of knowledge and theoretical practice, scientific explanation leads to the enrichment of existed knowledge of the world. For the other thing, in the result of scientific explanation, we could realize that why the antecedent theories did not successfully explain phenomenon which has problems. Scientific explanation also leads to the enrichment of meta-theoretical knowledge. The context of knowledge modification endows the cognitive process with explanatory meaning, makes it being explanatory act instead of descriptive or confirmative act. Just in this process, scientific explanation demonstrates vividly the contextual features of open-endedness, systematicness and cross-sectionness. Therefore, we argue that without certain context there is no such thing as scientific explanation. The plausibility, accessibility and acceptability of scientific explanation necessarily contain the conclusive role of contextual elements and features. In this sense, only the contextual analysis could clarify and specify the essential features and operational mechanism of scientific explanation.

IV. Summary

With considerable qualities of context-dependence and context-sensibility, the essence of scientific explanation is unfolded and accomplished in a contextualized dynamic process, an integration of contextual structure and elements. This integration contains the structurally unification of the syntactical basis, semantic rule, pragmatical boundary, and the fusion of explanatory question, background, subject and its intentionality. It is just in this combination, the explanatory regulation, process, criterion, and the diversity, multiplicity, and pluralism of the relevant model come into being. In conclusion, the contextualist idea could significantly enhance the comprehensiveness and open-endedness of scientific explanation theories. Leading to more broad multi-element notions of scientific theory and methodology, it also will encompass the intrinsic unitarity and common value in different thoughts of scientific explanation.

References

Cohen R. S. ed. (1993), *The Context of Explanation*, Kluwer Academic Publishers.

Caruana, L. (2006), *Science and Virtue: An Essay on the Impact of the Scientific Mentality on Moral Character*, Ashgate Publishing.

Dewey, John. (1910), *How We Think*, Lexington, Mass: D. C. Heath.

Gonzalez, W. J. (2010) *Explanation, Prediction, and Confirmation in the Social Sciences*, Journal of the General Philosophy of Science, Vol. 41.

Hempel, C. G. & Oppenheim, P. (1948), *Studies on the Logic of Explanation*, Philosophy of Science, Vol. 15, No. 2

Hempel, C. G. (1965), *Aspects of Scientific Explanation and other Essays in the Philosophy of Science*, New York: Basic Books.

Hesse, M. (1980), *Revolutions and Reconstructions in the Philosophy of Science*, London: The Harvester Press.

Kantorovich, A. (1988), *Philosophy of Science: From Justification to Explanation*, The British Journal for the Philosophy of Science, Vol. 39, No. 4.

Kitcher, P. & Salmon, W. C. eds. (1989), *Scientific Explanation*, Minnesota University Press.

Lipton, P. (2004), *Inference to the Best Explanation*, Routledge.

Potochnik, A. (2010), *Levels of Explanation Reconceived*, Philosophy of science, Vol. 77, No. 1.

Ruben, D.-H. (1990), *Explaining Explanation*, Routledge.

Salmon, W. S. (2006), *Four Decades of Scientific Explanation*, University of Pittsburgh Press.

Trout, J. D. (2007), *The Psychology of Scientific Explanation*, Philosophy Compass, Vol. 2/3.

Van Fraassen, (1980), *The Scientific Image*, Oxford: Clarendon Press.

ON A BOTTOM-UP APPROACH TO SCIENTIFIC DISCOVERY*

XIANG HUANG
School of Philosophy, Fudan University
No. 220, Handan Road, Shanghai, China, 200433
hangxiang@fudan.edu.cn

Two popular models of scientific discovery, abduction and the inference to the best explanation (IBE), presuppose that the reason for accepting a hypothetical explanation A comes from the epistemic and/or explanatory force manifested in the fact that observed fact C is an inferred consequence of A. However, not all discoveries take this top-down procedure from A to C, in which the result of discovery A implies the observed fact C. I contend that discovery can be modeled as a bottom-up procedure based on inductive and analogical rules that lead us to infer from C to A. I take the theory of Dignaga, an Indian medieval logician, as a model of this bottom-up approach. My argument has three panels: 1) this bottom-up approach applies to both commonsense and scientific discovery without the assumption that C has to be an inferred consequence of A; 2) this bottom-up approach helps us get around problems that crop up in applying abduction and/or IBE, which means that scientific discovery need not to be modeled exclusively by top-down approaches; and 3) the existence of the bottom-up approach requires a pluralist attitude towards modeling of scientific discovery.

I. Introduction

Scientific investigations aim to find hypothetical explanations to account for the facts obtained as a result of direct observation and designed experiments. The process of finding these hypothetical explanations is scientific discovery. Two popular models of scientific discovery in contemporary philosophy of science are abduction and its variation, the so-called "inference to the best explanation" (i.e. IBE). Both of them try to grasp the rational and the epistemic dimension of the non-deductive character of inference from observed facts to hypothetical explanations. They originate in Charles S. Peirce's insight: "the surprising fact, C, is observed; But if A were true, C would be a matter of course; Hence, there is reason to suspect that A is true". [a]

In both abduction and IBE, it is the fact that C features as an inferred consequence of A (or $A_1 \cdots A_n$), which gives a reason to think that A (or $A_1 \cdots A_n$) might be an explanation of C, and the explanatory power embedded in this infer-

*This research has been sponsored by National Funds of Social Science (11BZX022), Shanghai Pujiang Program, 2011RWXKZD013 and CONACYT CB2009/133345.

[a]It is the last of various forms that Peirce developed in his life. For the development of different forms and their subtle differences, please see Niiniluoto's[1] and Aliseda's[2]work.

ential relation is supposed to lead us to find and believe A (or $A_1 \cdots A_n$)'s being true.[b] I call this way of modeling scientific discovery the top-down approach, as it takes the observed fact C as the inferred consequence of the result of discovery, instead of inferring the result of discovery A (or $A_1 \cdots A_n$) from C. That is, in the top-down approach, it is the fact that C features as an inferred consequence of A (or $A_1 \cdots A_n$) that makes the discovery possible, which is not the same as inferring the result of discovery from C. One apparent problem of the top-down approach comes from the fact that, in an inference assumed to model the process of discovering the hypothesis A (or the hypotheses $A_1 \cdots A_n$ individually) from the observed fact C, the result of discovery A (or $A_1 \cdots A_n$) is already included in the premise expressed by the formula, "A (or $A_1 \cdots A_n$ individually) implies C". The appearance of the result of discovery in the premise of an abductive inference makes some scholars doubt whether Peirce's inference is truly a model of discovery, e.g.[4] In the case of IBE, the fact that inquirer should already be in possession of possible explanations ($A_1 \cdots A_n$), among which she compares, has cast doubt on the claim that IBE is a discovery procedure, since it does not provide a sufficient explanation of how ($A_1 \cdots A_n$) are generated. Rather, they are the givens from which IBE starts.[5]

If the result of discovery is presupposed in the premise of the model of discovery, this does not necessarily refute the top-down approach. Defenders of abduction and IBE can deal with this problem in different ways. For example, Ilkka Niiniluoto[1] argues that abduction and IBE might be taken as reliable heuristics whose rational force is articulated in the Bayesian reconstruction of Peirce's schema of abduction. Jaakko Hintikka[6] and Sami Paavola[5] contend that logics of discovery should be embedded in the context of interrogative argumentation instead of being considered as algorithmic rules in a particular system of logic. Rather, they are strategic rules that suggest plausible ways of finding hypothesis. I will not take sides in this controversy, because I want to tackle the problem of presupposing the result of discovery in the premise by suggesting an alternative model of scientific discovery. In the next section, I will elaborate this alternative as what I call a bottom-up approach, in which the observed fact C need not feature as an inferred consequence of the explanatory hypothesis A, but as a reason to supporting it. In the third section, I show that this alternative can also explain the case of Ignaz Semmelweis' discovery of the cause of childbed fever, which has been canonically referenced to support different top-down approaches. In the fourth and the fifth section, I compare this bottom-up approach with abduction and IBE, arguing that it is autonomous and epistemically irreducible to either mode. Finally, in the conclusion, I defend a pluralism with regard to scientific discovery, according to which different inquiry contexts and different types of problems require different kinds of discovery approaches. The

[b]This aspect is quite clear in the second premise of Peirces abduction. In Peter Liptons IBE, this aspect can be seen in this description: "[g]iven our data and our background beliefs, we infer what would, if true, provide the best of the competing explanations we can generate of those data (so long as the best is good enough for us to make any inference at all".[3]

bottom-up approach is just one of the set of such approaches. This pluralistic view gives us a better description of the way the logics of scientific discovery unfold in the world than what the top-down approaches give us.

II. A Bottom-Up Approach: Dignaga's Triplet (DT)

Finding a hypothetical explanation A for a surprising observed fact C is a cognitive process that happens both in scientific and commonsense practices. In ordinary life, the cognitive process of finding a possible explanation for a puzzling fact does not proceed through logical implication, even though the process can be reconstructed later using logical notions. You see the carpet is stained with ink, and you guess immediately that the inkbottle was knocked over. Even though the process can be easily reconstructed with logical relations, you need not make the logical implication to guess what happens. Before the western notion of logical implication was discussed in the Non-Western world, the above mentioned cognitive practices had been the standard for commonsense thinking, extensively used not only in ordinary life but also in scientific and technological practices.

In medieval India, before a theory of reasoning equivalent to the western formal logic or the syllogism was developed around the 7th and 8th century, systematic studies of the reasoning that inferred from the observed fact C to a hypothetical explanation A were common. Two reasons might be responsible for medieval India's lack of formal logic. First, medieval Indian thinkers often located their theories of reasoning in the context of argumentation, assuming that norms of debates, considered by Western logics as purely rhetoric elements, were not reducible to formal treatment. Second, inferences normally concerned only with particular objects and situations, since there was no notion, as yet, of universal statement; and, as a consequence, all inferences were essentially analogical.[7] As we will see, this second aspect is particularly relevant to our argument, because it suggests how we usually infer a hypothetical explanation A from the observed fact C without assuming that C is an inferred consequence of A.

The theory of reasoning I will use to illustrate this bottom-up approach comes from the Buddhist philosopher, Dignaga (circa 400-480). This is not an exercise in intellectual history, but instead aims to grasp the main spirit of the bottom-up approach. In a narrow sense, Dignaga's theory of reasoning is a theory of inference (*anumana*), whose scope comprises "besides inferences of everyday life and proofs of particular doctrinal tenets also the establishment of what might be called 'scientific theorems' ".[8] In a broad sense, it is also an epistemic theory about how we acquire knowledge (*pramanas*), which takes it for granted that correct inferences are reliable means of knowing. For Dignaga, there are only two ways of acquiring knowledge: directly from experience and indirectly from reason. The inference from an observed fact C to its hypothetical explanation A falls under the latter. A standard example among the Indian philosophers was that of inferring fire from smoke. Suppose we see smoke billowing upwards on a hillside (C), then we infer that the hill is on fire (A). Dignaga's theory of reasoning provided three rules to grasp the normative and epistemic dimension of this kind of inference. I propose to call them "Dignaga's Triplet (DT" According to Richard Hayes, one of the great specialists on Dignaga, DT can be articulated with western philosophical notions in this way:

(1). The inferential sign must be a property of the subject of the inference. That is, there exists in the subject of inference a property, which is different from the inferable property and which is furthermore evident to the person drawing the inference; this second property may serve as an inferential sign in case it has two further characteristics.
(2). The inferential sign must be known to occur in at least one locus, other than the subject of inference, in which the inferable property occurs.
(3). The inferential sign must not be known to occur in any other loci in which the inferable property is absent.[20]

In order to understand the relevance of the DT to our topic of discovery, we need to reinterpret its technical terms within our context of inferring a hypothetical explanation A for an observed fact C. In the case of the hill on fire, A is the claim that the hill is on fire, C is observed smoke on the hillside. In the first sentence of Hayes' (1), "the inference" refers to A, which is the hypothetical explanation inferred; "the inferential sign" refers to C, which is the observed fact that features as a sign or a possible reason for inferring A; "the subject of the inference" refers to the hill. The whole sentence says that observed smoke (C), must be the property of the hill in some state (A). In the terminology of contemporary philosophy, this sentence requires that the observed smoke must be indeed present on the hill to which the hypothetical explanation refers; that is, it can't be fake or hallucinatory smoke, or smoke observed on other hills. The second sentence in (1) says that the hill, or the subject of the inference, must have a property of fieriness, which (a) is a property different from the property of the observed fact C, which is the smokiness observed on the hill, and (b) is evident for the person who wants to infer A from C. Only then, it is claimed, may C serve as a sign or a reason for A if it also satisfies other two conditions (2) and (3).

The condition (2) requires that the positive example of the property appeared in C, which is the observed smokiness, should have appeared at least once in a place in which the property of A, fieriness, also appears; however, this appearance of fieriness in the positive example should not be identical with the case under observation. We must have some sense of fire and of smoke at places other than where the hill we are observing locates. A suitable analogy to smokiness as a property of fire can be found in kitchens, where the occurrence of the properties of smokiness and fieriness can appear together. Furthermore, we know that a kitchen is not the subject of A, which is the hill. It is easy to see that the requirement of having at least one positive example is asking for evidential support of A out of a range of experiences in which the association of smokiness and fieriness has been affirmed, such as in a culinary hearth. However, there is something obscure about the requirement that the positive example should not take place at the place in which the subject in discussion happens. However, if the evidential support, in our case, the association between smokiness and fieriness, takes place uniquely in the same hill that A references, this evidential support would be *ad hoc* in character. Essentially, it would make the association between smokiness and fieriness in the hill the explanation for the association between smokiness and fieriness in the hill. Condition (2) is an epistemic consideration, and its way of excluding the *ad hoc* character in evidential support is about the epistemic condition of the observation. I will return to this problem in the section 4.

The condition (3) requires that in a place in which the property of A, or the fieriness in our case, is absent, the inferential sign C, the smokiness, has not been observed. If, for instance, the observer were looking at a lake, (where fieriness is absent) and doesn't see smoke, we have a negative instance of non-fieriness associated with non-smoke. It is to be observed that, due to the absence of universal quantifiers in the logic with which Dignaga was familiar, (3) is not a universal negative proposition, claiming that anything that is not-fiery is not smoky. On the contrary, (3) should be understood as an inductive observation: we simply have not observed a place in which the fieriness is absent but smokiness is nevertheless present. According to Dignaga, once an inferential sign C has satisfied all the three conditions, C can be a reliable or trustworthy reason to infer A, which is an explanation of C itself.

With the help of the case of the smoking hill, we see how we can infer a hypothetical explanation A from C. In fact, Hayes' expression of DT can be reconstructed with the Peircean terms; that is, we can reasonably infer a hypothetical explanation A of the observed fact C, if the inference satisfies the following three conditions:

(DT-1) For the person who draws the inference, the subject of the hypothetical explanation A indeed has the property of the observed fact C, that is, A is indeed applicable to the explanation of C;
(DT-2) The property of C must appear at least once in a place in which the property of A appears, but what occurs in this place is not the subject of A;
(DT-3) We haven't observed a place in which the property of A is absent but the property C is observed.

DT can be easily applied to many cases of discovery processes in ordinary life. For example, you open the window and see that the lawn is very wet, C, from which you infer that it has rained, A. If someone asks you why you think it poured rain, of course, you can say that heavy rainfall implies getting the lawn wet. But if she continues to ask why a soaking lawn cannot have become wet through other causes, she is asking you to give her a reason to believe that your inference is reliable. The Peircean abduction aims to offer an explanation of the reliability. Alternatively, DT offers another, according to which this inference is reliable when it satisfies the following conditions: for you, the lawn puddled with rainwater is the lawn that you are seeing (DT-1); you can allude to other places where this degree of wetness is the result of rain, as for example a football field soaked after a rain (DT-2); and, finally, you haven't encountered a situation in which the lawn is soaked without the rain soaking it. For example, a fine mist doesn't visibly leave the lawn soaking wet (DT-3).

DT has two interesting aspects. Firstly, as a model of discovery in which explanatory hypotheses are inferred from observed facts, DT is logically different from abduction and IBE: it does not take the assumption that C is an inferred consequence of A as a premise of its inference. Rather, as (DT-3) presents us with an inductive and fallible association, DT denies that C is a logical consequence of A. In this sense we can say that DT has to be a bottom-up approach of discovery, contrary to the top-down approach exemplified by abduction and IBE. As we have seen, people from different parts of the world were already in the habit of inferring hypothetical explanation from observed facts long before the notion of logical consequence had been formalized in the West, which supports our claim that bottom-up approaches such as DT are the more natural way of modeling the cognitive process of discovery.

The second aspect of DT that interests us is its analogical character. DT tries to provide a criterion for having a reliable inference from C to A. The normative force of the criterion is not centered on C being necessarily an inferred consequence of A, as the top-down approach assumes, but from the three rules we have looked at above, which C should satisfy when it relates with A. As we have seen with (DT-3), the relation between C and A is inductive. After satisfying the requirement in (DT-2), putting together in one place the appearance of C and the property of A, (DT-3) requires, more strongly, that when the property of A is absent, C does not appear. This make DT analogical in character, as (DT-2) asks us to find at least one positive example of which the similar explanatory relation between A and C has been appeared in other occasions; and the requirement in (DT-3) asks us to search for any negative example that can refute this relation. The reliability of inferring A from C comes from the practice of searching for these particular examples and from careful comparisons of the similarity and dissimilarity among them. The analogical character is quite significant for the bottom-up approaches of discovery. In fact, the analogical examples in many cases help to conceive A as a possible explanatory hypothesis, as we will see in Semmelweis' case in the next section. This means that DT models aspects of generative procedures in discovery. Hence, while the normative force of the top-down approaches comes at least partially from the logical relation between A and C, bottom-up approaches are based on the analogical relations grounding the inference of A from C.

III. DT and Semmelweis' Discovery

We have seen that DT can model commonsense inference from an observed fact to its hypothetical explanation. In this section, I will show that DT can also model scientific discovery. The example that I will use is Ignaz Semmelweis' investigation of the cause of childbed fever. This case, which is famous in the history of medicine, is also famous in the history of the philosophy of science, where it was introduced as a model for understanding scientific discovery by Carl G. Hempel. Hempel theorized that creative processes of inventing hypotheses in science cannot be routinized according to any logic, and that the objective validation of these hypotheses should be modeled by the hypothetic-deductive method. For Hempel, Semmelweis' case exemplifies the hypothetic-deductive method.[9] Peter Lipton sees the example differently, understanding Semmelweis' investigation under the aegis of IBE; that is, for Lipton, Semmelweis chose the best explanation among several alternatives.[3] Recently, Paavola argued that Semmelweis did not actually find the best explanation through eliminating competing hypotheses, but rather came to his conclusions through Hansonian abductive inquiry. Semmelweis "did not produce any 'pool' of candidates from which he would have eliminated the candidates. Rather, he generated alternative explanations, which he tested and evaluated one by one until he found an explanation that worked".[10]

I agree with Lipton and Paavola that there are some patterns of discovery that are best understood as epistemic processes, Semmelweis' case being one of them. I also agree with Paavola's idea that Lipton's reconstruction by IBE is too artificial, since Semmelweis did not first generate a pool of candidate explanations and then choose the best among them. But I disagree that Semmelweis engaged in something like Hansonian abduction in explaining the case. According to Paavola, the Hansonian abduction manifested in Semmelweis' case is "an inference from some constellation of curious or characteristic features (with a proviso that these features would follow from some hypothesis) to a conclusion that this kind of a hypothesis is a good candidate explanation in this specific case. In this sense, in typical cases, abduction is reasoning backwards from some effects to their potential causes, or to the potential type of a hypothesis".[10] However, "reasoning backwards from some effects to their potential causes, or to the potential type of a hypothesis" need not follow the top-down pattern, by which I mean that the inquirer need not suppose that curious or characteristic features "would follow from some hypothesis"; on the contrary, if DT is sustainable in Semmelweis' case, it can be a bottom-up type that infers from these features to their hypothesis, not with the proviso that these features logically follow from the hypothesis, but with some analogical considerations to establish a rational relation between these features and the hypothesis.

Even though Lipton's interpretation of Semmelweis' case by IBE is an artificial reconstruction, his illuminative classification of the possible hypotheses that Semmelweis had considered does help us to understand the nature of these hypotheses. Semmelweis wanted to explain an observed fact C, to wit, that at the Vienna General Hospital, the First Maternity Division had a higher proportion of childbed fever than the Second Division. In order to explain it, from 1844 to 1848, he considered a series of possible hypotheses. Lipton identifies three types of these hypotheses: (1) hypotheses that could not differentiate between the clinics, and so were rejected; (2) hypotheses that could differentiate, but where the elimination of the putative cause did not affect the difference in mortality; and (3) hypotheses that differentiated and in which the elimination of the putative cause eliminated the difference in mortality.[3] The last type was capable of explaining the high proportion of childbed fever at the First Division, and Semmelweis finally found that doctors and medical students bringing cadaveric matter into contact with the mothers' bloodstream satisfied this type of hypothesis. From DT's perspective, the cadaveric hypothesis satisfies DT. In this case, A is the cadaveric hypothesis, according to which the cadaveric matter introduced by doctors and medical students working at the First Division was associated with the high proportion of childbed fever; and C is the observed fact that the proportion of childbed fever at the First Division was systematically higher than that of the Second Division. The cadaveric hypothesis satisfies (DT-1) because, for Semmelweis, the person who drew the inference from C to A, the cadaveric matter that is the subject of A was indeed introduced to the First Division, which is the central focus of C. (DT-2) requires that there is at least one external situation in which A occurs and the property C is present. This example was Semmelweis' colleague, Kolletschka, who got his blood contaminated with "cadaveric matter" (the property of A) during an autopsy, and died from an illness with symptoms similar to that of childbed fever (the property of C). In fact, it was this unfortunate death that prompted Semmelweis to think about the danger posed by cadaveric matter. (DT-3) is the requirement that in a place where the property of A is absent, the property of C also disappears. This corresponds to the fact that Semmelweis disinfected the hands of the medical students in a solution of chlorinated lime before examination (to let the property A disappear), and that the mortality rate in the First Division lowered to the same level as that in the Second Division (that is, C disappeared).

The second type of hypotheses differentiated the two, yet the mortality in the First Division occurred at the same rate even after the putative cause had been eliminated. Semmelweis had gone through the factors that might be particular for the First Division, such as overcrowding, religious practices, the delivery position adopted in the First, but were not present in the Second Division, etc. But if we control for all these factors in both divisions, the mortality rate in the First Division still remained higher than that of the Second. From DT's point of view, this type of hypotheses violates both (DT-1) and (DT-3). It violates (DT-1), because after eliminating the putative cause of the difference, it turns out that the higher mortality in the First Division could not be attributed to these hypothetical factors. In the terminology of DT, the subject of A cannot have the property of C. This kind of hypothesis also violates (DT-3), for the elimination of the putative cause was an occasion in which the property of A was absent, but the property of C - the higher rate of mortality - was still present.

The first type of hypothesis did not differentiate between the two divisions. For instance, take the hypothesis that some kind of epidemic disease was responsible for the higher rate of mortality in the First Division. Semmelweis quietly rejected it because it failed to explain why the mothers of the First Division were more prone to suffer from childbed fever, since there was a random chance of the mothers being allotted to the first or the second division. From DT's point of view, this hypothesis violates (DT-2) that requires that there is at least one example in which both the property of A and the property of C appear together in some places other than the place that the subject of A occurs. In the case of the hypothesis of epidemic disease, (DT-2) requires that the same kind of disease with a high rate of childbed fever should appear in some other place than the First Division. However, Semmelweis could not find such a place. On the contrary, he found that in a place vulnerable to the diseasethe street on which some mothers gave birth before they could arrive at hospitalthe rate of childbed fever was not high. That was a case in which the property of A appeared but C didn't. [c]

IV. DT's Irreducible Epistemic Aspects I

The fact that DT is applicable to the commonsense and scientific discoveries from an observed fact C to a hypothetical explanation A does not directly imply that it is an appropriate way of modeling scientific discovery. We have already seen that DT does not require that C is an inferred consequence of A, but only specifies three rules for having a correct inference from C to A. One may object that if these three rules can be reconstructed by and hence reducible to other models of discovery, such as abduction and IBE, then DT could be plausibly construed as merely a primitive intuition of discovery practices that were systematized and given formal expression by the invention of the Western logics. In this section, I argue that this reductionist account doesn't work, because the epistemic dimension of DT goes far beyond that of abduction and IBE. Four reasons will be presented: 1) DT's analogical character, which is quite useful for explaining how a possible hypothesis can be generated, is not reducible to abduction or IBE; 2) due to a stricter normative requirement, DT is capable of avoiding the so called "wild" hypotheses problem suffered by abduc-

[c]We can have hypotheses that combine the aspects of different types. In 1846, Semmelweis also considered a hypothesis that "the prevalence of illness in the First Division to injuries resulting from rough examination by the medical students, all of whom received their obstetrical training in the First Division. Semmelweis notes in refutation of this view that (a) the injuries resulting naturally from the process of birth are much more extensive than those that might be caused by rough examination; (b) the midwives who received their training in the Second Division examined their patients in much the same manner but without the same ill effects; (c) when, in response to the commission's report, the number of medical students was halved and their examinations of the women were reduced to a minimum, the mortality, after a brief decline, rose to higher levels than ever before".[9] This was a hypothesis that at first glance pertained to the third type and seemed quite attractive. However, (a) and (b) converted it into the first type, and (c) converted it into the second; hence, Semmelweis finally rejected it.

tion; 3) DT requires that ad-hoc hypotheses should not be taken into consideration for explaining C, which is not required from abduction and IBE; and 4) one strong reason for adopting IBE as a model of scientific discovery is that scientists may use A's explanatory power as a guide for evaluating A. However, I claim that DT is not less suitable for being a guide, though with different normative rules. I will examine the first three reasons that concern with the differences between DT and abduction in this section, and leave 4) to the next section. I hope that these comparisons may help us to understand better the nature of DT and of its epistemic status.

It is commonly said that a process of discovery consists of three cognitive stages: a) first the generative stage, in which a candidate hypothesis is founded in order to explain a phenomenon or a fact in question; b) second the evaluative stage, in which the acceptability of a candidate hypothesis is established; and c) thirdly the selective stage, in which a best hypothesis is selected among alternative candidate hypotheses. Abduction concentrates mainly on a) and b); whereas IBE concerns basically b) and c). As Nickles points out, these three stages need not be temporally ordered and may happen in different order in different scientific inquiries. The generative stage is particular problematic for abduction, as an abductive inference has only two parts, the observed fact C, and the logical relation between C and it possible hypothesis A, without further explanation about where A comes from. This has disturbed some philosophers, who question the legitimacy of abduction as an adequate model of discovery.[4]In comparison with abduction, DT displays a generative procedure which, if not complete, is at least intuitive. As we have seen, Semmelweis begin to form the cadaveric hypothesis in response to the unfortunate death of Kolletschka, which is precisely the scenario envisioned by (DT-2), which asks for, an example of a place in which both the property of A and the property C appear. This example is analogical, and depends on the observer recognizing a similarity between cases here, Semmelweis seeing the similarity between the symptom of Kolletschka's illness and that of childbed fever. This similarity led Semmelweis to think for the first time that the infectious material which was handled by medical students doing autopsies might be the vector by which both Kolletschka became ill and the mothers suffered childbed fever. However, we have no reason to think that (DT-2) can be reduced to this computational model, because (DT-2) has more epistemic elements than are allowed in Thagards computational model, which blocks the reduction of DT to abduction in all cases.[d]

The second reason that DT is not reducible to abduction and IBE is that DT's normative rules in evaluative stage are in a sense more restrictive than that of abduction, because they rule out the problem of "wild" hypotheses that notoriously

[d]The analogical character of DT implies that DT can be considered as an inductive reasoning from particular to particular. In fact, most specialists on Dignaga recognize the inductive character of his reasoning theory. If DT can be established as an model of scientific discovery, it implies that, even if Hempel is right in claiming that there areno generally applicable "rule of induction", some locally applicable rules of induction, such as DT, do let us mechanically infer hypotheses from empirical data.

dogs abduction centered accounts of discovery. Peter Achinstein[11] expresses the classic statement of the problem like this: "the hypothesis that I will be paid one million dollars when this paper is published would, if true, explain why I am writing the paper. But this provides no reason for thinking that I am about to become a millionaire. There are many 'wild' hypotheses which if true would explain the date, but unless there is some other evidence in their favor this fact by itself leads no plausibility to them". This problem leads Achinstein to conclude that there is no difference between the logic of discovery and the logic of justification. DT is, of course, more than a "logic" of discovery, for as we have seen it displaces the focus of the process from a logic to the epistemic conditions of analogy. Taking Achinstein's example from the DTs perspective, the procedure we would follow would go like this. First, the observed fact C is that I'm writing a paper, and the hypothetical explanation A is that I will be paid one million dollars for its publication. The inference from C to A satisfies (DT-1), that is, the fact of writing the paper C is precisely what the subject of A is doing. However, it is not clear whether the inference can satisfy (DT-2), since as we cast about we find that it is very rare that an academic paper in philosophy earns one million dollars—in fact, no instance comes to mind. This is pertinent, in so far as it is I who am writing the paper, and whose motives are being reconstructed—and it is I, as well, who have found no instance of a philosophy paper earning its author a million dollars. Finally, what definitely makes this inference ineligible is (DT-3), according to which, in a place where the property of A is absent, the occurrence of the property of C hasn't been observed; that is, (DT-3) requires that in the situation that I am not offered one million dollar for an article, I would not write a paper for publication. But, in fact, although I have never been offered one million dollars to write an article, I have written and published many academic papers. In this way the wild hypothesis would be rejected by DT as invalid. The problem of wild hypotheses is a consequence of abduction's assumption that C's being inferred consequence of A gives a reason to believe that A is a hypothetical explanation of C; however, C might be logical consequence of any one or more of an unlimited number of wild hypotheses. DT avoids this problem with the strict restriction from (DT-2) and (DT-3). This particular epistemic dimension of DT offers another evidence for thinking that DT cannot be reduced to abduction.

The third reason that DT is irreducible to the logical modes of scientific discovery comes from the way it avoids *ad hoc* hypotheses. As mentioned in the section 2, (DT-2) requires that the positive example of the property of C and the property of A appear together should not happen in a place in which the subject of A occurs. This requirement is imposed to avoid *ad hoc* explanations. This is an important epistemic point worthy of a more detailed consideration. To begin with, we need a clarification of the *ad hoc* explanation. When Hempel and Oppenheim[12] suggested the Deductive-Nomonlogical (D-N) model of explanation in 1948, they proposed three logical conditions of an adequate explanation:

R1: The *explanandum* must be a logical consequence of the *explanans*.
R2: The *explanans* must contain general laws, and they must be essential for the derivation of the *explanandum*.
R3: The *explanans* must have empirical content; that is, it must be capable, at least in principle, of test by experiment or observation.

Critics have observed that R3 is in a sense can be redundant because, if the *explanandum* is a logical consequence of the *explanans* (R1), the same *explanandum* is an empirical component of the *explanans*, hence R3 must necessarily always be satisfied.[e] However, this misunderstands Hempel and Oppenheim's point. As the discussion in their article of 1948 shows, the empirical content required in R3 refers exclusively to the testable evidence for the *explanans* other than the *explanandum*. An explanation with explanans whose only empirical content comes from its *explanandum* is a situation in which the reference set of the candidate hypothesis A coincides with the reference set of the *explanandum* C, which is just what an *ad hoc* explanation is. The requirement "what occurs in this place is not the subject of A" in (DT-2) is a way, using our method, that explicitly excludes this situation for similar reasons.

However, DT does not simply retranscribed Hempel's type of explanation. Not all explanations that satisfy the D-N model of explanation are acceptable in DT. For instance, we can explain why human beings are thinking beings C, by adopting the candidate hypothesis A, that this is because human beings are rational animals. In DT, the candidate hypothesis A cannot be reasonably supported by C, precisely because C does not satisfy the requirement "what occurs in this place is not the subject of A" in (DT-2), as you cannot find a place in which being an rational animal (the property of A) and being an thinking being (the property of C) appear together outside the category of human being. Since all rational animals are human beings, you cannot find an analogical example outside the subject of A, which covers all category of human being. However, this explanation A can still be considered as an elliptically formulated explanation of Hempel's type, reconstructed by the following valid syllogism:

Rational animals are thinking beings;
Human beings are rational animals;
Therefore, human beings are thinking beings.

This syllogism satisfies all three logical conditions of adequacy of the D-N model, that is, the conclusion is a logical consequence of the premises; the first premise, the

[e]The similar problem is also suffered by the interpretation of the second premise $A \to C$ as A explains C in the Peircean abduction. As Aliseda observes: "As for the testability requirement, when the second premise is interpreted as logical entailment this requirement is trivialized, since given that C is true, in the simplest sense of 'testable', A will always be testable".[2]

elliptical one, can be considered as a general laws; and, all premises have empirical content by pointing out that some individual human being are both thinking beings or rational animals. According to Hempel and Oppenheim, A must be a good explanation as it satisfied the D-N model; but, according to DT, it is a bad one as it is ad hoc, violating (DT-2).

The difference between what is validated by these two different methods of discovery is very interesting. In D-N model, as in abduction, the explanatory relation between A and C is understood as a logical relation in which A logically implies C. The evidence for the premises is precisely the same evidence for the conclusion. For example, the evidence for the first premise is the person who is capable of thinking, and it is also the evidence for the conclusion. This evidential support gives an epistemic force to both the D-N model and abduction, though in different ways. In DT, where the logical relation between A and C is not assumed, the inference from C to A is only analogical, and the epistemic force of it is quite different. What evidentially supports A cannot be totally equivalent to that which supports C. The ad-hocness is transgresses against the DT procedure of drawing analogical example that may always have extra empirical content other than what C has. The extra empirical content is important to the property of novelty, or discovery, which is what the procedure is about. The generative stage of having A and the analogical inference from C to A exists to validate an innovation—a discovery. The hypothetical explanation A must, thus, play the important epistemic role of being novel (this is, not something that can be deduced from its logical structure) which is separate from its role as providing evidential support.[f] Hence, the urgency of avoiding the *ad hoc* explanation is what the analogical character of DT stresses.

V. DT's Irreducible Rpistemic Aspects II

The fourth reason that DT is irreducible to abduction or IBE comes from its way of using the explanatory power of A in its evaluation. At this point, the difference we want to highlight is between DT and IBE, as the latter is constructed around a heavy emphasis on the evaluative stage (inherent in the notion of "better" or "best"). While the evaluative stage of abduction is concerned with the plausibility of a hypothetical explanation A, IBE's evaluative stage always constructs alternative hypothesis $A_1 \cdots A_n$ among which to chose the one with the greatest and most applicable explanation power. I take Lipton's IBE as an example since it is

[f]As Lakatos points it out, "a bold theory should have some novel potential falsifiers which none of the theories in the extant body of science has had", and, "it is excess empirical content, that is, the increase in empirical content, rather than empirical content as such, that measures the boldness of the theory".[13] The innovation and the boldness together form an important consideration of the initial plausibility of the candidate hypothesis, which is the starting point of the evaluative stage. Besides, they are also important for the selective stage, as they affect the decision about which candidate hypothesis is better.

one of the most sophisticated theory of its kind. [g]Lipton does not take the evaluative stage and the selective stage as two consecutive processes; on the contrary, he thinks that when scientific discovery begins, the selective considerations have already been brought into the evaluative process. More specifically, the explanatory power of a possible hypothesis is our guide to evaluating the epistemic status of this hypothesis. It is true that explanatory power is one of the essential criteria for evaluating hypothesis. For example, for Semmelweis, one reason to accept the cadaveric hypothesis was that it could not only explain the high mortality in the First Division, but also explain "the lower mortality among 'street birth': women who arrived with babies in arms were rarely examined after admission and thus had a better chance of escaping infection".[9] Lipton's IBE grasps very well the spirit of these considerations. However, there are at least two difficulties that his IBE must be questioned about. Firstly, not all discovery processes begin with the candidate hypotheses already in place. In the case of Semmelweis, as Paavola has observes, "[Semmelweis] did not produce any 'pool' of candidates from which he would have eliminated the candidates. Rather, he generated alternative explanations, which he tested and evaluated one by one until he found an explanation that worked".[10] If Paavola's observation is right, then IBE cannot model all types of scientific discovery as Lipton thinks, but only some of them; and then at a certain stage in the process. Hence, we do not have any *a priori* reason to think that DT should be reduced to IBE.

The second difficulty is that Lipton's notion of explanatory power is too vague as it stands. Extra epistemic resources need to be added to get it going. However, the strategy of repairing the epistemic weight of explanatory power is not an easy task. As I will argue, sometimes DT can assist us here. In Lipton's IBE, the explanatory power is partially characterized by a series of elements that deal with understanding, such as scope of the hypothesis, precision, mechanism, unification, simplicity, etc. These elements make a hypothesis lovely, or easy to be understood. However, sometimes it is difficult to give a unique definition of the loveliness of the hypothesis because many of these elements gain their stature from their implication in different kinds of psychological dimensions. Another aspect of explanatory power is its likeliness. It concerns with the truth of the explanation, which is an

[g]Some IBE theories are easily identified as a kind of the top-down approach, such as Ben-Menahem,[14] Day and Kincaid,[15] Fetzer,[16] etc., which take the logical or probabilistic relation between C and every $A_1 \cdots A_n$ separately as a necessary factor to measure the explanatory power of the latter. Lipton's IBE is more complicated. His analysis of explanatory loveliness clearly goes beyond the scope of the top-down approach. However, I think I can still consider Lipton's IBE as a kind of top-down approach, because it characterizes of the likeliness of a hypothesis based on the evidential and inferential relation between C and every $A_1 \cdots A_n$ separately, which when then measure with Bayesian tools. Admittedly, this is an elegant mode of explanation that certainly increases the scope and sophistication of the relation. Lipton argues that his explanationism and Wesley Salmon's Bayesianism should be complementary, because "explanatory considerations may help enquirers to determine prior probabilities, to move prior to posterior probabilities, and to determine which data are relevant to the hypothesis under investigation".[17]

epistemic consideration very different from the loveliness. Lipton introduces Mill's method of differentiating to model the contrastive explanation of why P rather than Q happens. Combining together the two different components, loveliness and likeliness, sometimes makes Lipton's picture of the discovery processes quite artificial. To return to the Semmelweis example, which Lipton discusses too, Lipton makes great effort to explain why Semmelweis rejected the first type of hypothesis that could not distinguish between the clinics. Remember DT gives a straightforward answer to it: this type of hypothesis violates (DT-2), as Semmelweis could not find a place other than the First Division that had both the property of A and that of C appeared together. In fact, Mill's method also gives a straightforward answer: an aspect that appears at both the place in which the effect is produced and the place in which the effect is absent cannot cause the effect. Here, no alternative candidate explanations are under consideration. However, in order to put this consideration into a contrastive situation for different candidate hypotheses, Lipton has to complicate Semmelweis' consideration.

In order to give a place for the consideration of candidate hypotheses, Lipton maintains that Mill's straightforward answer is not sufficient, because this answer cannot reject the hypothesis of the first type in Semmelweis' case, but only shows that the hypothesis fails to explain some evidence, but "no hypothesis explains every observation, and most evidence that is not explained by a hypothesis is simply irrelevant to it".[3] According to Lipton, to really reject the first type of hypothesis, we need to take into account the alternative hypothetical explanations; and these alternative hypothetical explanations have two kinds of relations with the original hypothesis: they either contradict the original hypothesis, or substitute for it a more complete explanation, which shows it to be unnecessary "so that the evidence that originally supported it will instead support a more complete competitor that can explain more phenomena". Lipton gives an example of the substitution of a more complete hypothetical explanation: "suppose that we had two wards of patients who suffer from syphilis and discovered that many more of them in one ward contracted paresis than in the other. The hypothesis that syphilis is a necessary cause of paresis would not explain this contrast, but this would notlead us to abandon the hypothesis on the grounds that its support had been undermined".[3] By making the distinction between the two kinds of relations, Lipton puts the discussion of the first type of hypotheses into the familiar context of the competition among rival hypotheses, driving us towards IBE. He says: "so the mechanism here turns out to be an attenuated version of disconfirmation by rival support after all", and "on this view, given a hypothesis to explain a contrast between that fact and a similar foil, the scientist must choose between the overarching explanations that the failure is due to incompleteness or that it is due to incorrectness".[3] Semmelweis rejected the hypothesis of the epistemic disease regarding it as incorrect, but Lipton believes that this rejection might be motivated by the desire for a more complete hypothetical explanation. For Lipton, even the cadaveric hypothesis might not be a complete explanation, because it could not explain why, in both the First and the Second Division, some mothers suffered the childbed fever and others not, why some mothers in other hospitals suffered the fever even though they were not examined by the medical students in the First Division.

DT suggests two observations on Lipton's treatment of the first type of hypotheses in Semmelweis' case. First, during the analysis of this topic, Lipton seems to drift from the object of explanation C, that is, from "why the First Division had a higher mortality than the Second", to another question, "what is the cause of the childbed fever". These are evidently two different questions. The answer to the first may or may not help you to answer the second, but is independent of it; whereas the answer to the second is at least going to be a factor in the final answer to the first. Semmelweis' investigation was toward the first, and it was successful in bringing down childbed mortality without developing an etiological explanation of childbed mortality, the answer to the second question. This could only be given later, after the development of germ theory, the improved understanding of the physiology of the immune system, etc. Mixing up these two questions may introduce easily "overarching explanations" in an IBE style. But it was legitimate for Semmelweis to try to answer a practical, local question: "why did the First Division have a higher mortality than the Second?" His cadaveric hypothesis, as it turns out, needed not to explain the phenomena beyond a direct answer to C.

The second observation is that DT gives a clear distinction between the explanation referring strictly to syphilis as the ground of paresis and the epidemic hypothesis. The hypothesis of having syphilis A explains the fact that there are patients who suffer the paresis C_1, because the inference from C to A satisfies DT. The hypothesis of having syphilis alone, however, does not explain the fact that paresis is not always present in syphilis patients. That variegated aspect of syphilis might be labeled C_2. C_2 is clearly not C_1, and thus, the inference from C_2 to A violates (DT-1), as the subject of A refers to all patients of syphilis, and the property of C_2 is applicable to some particular members of them. In order to explain C_2, there we need a new hypothesis A according to which, the syphilis patient that haven't been treated early are more likely to have paresis. Semmelweis' object of investigation was similar to C_2. Shifting the object of explanation from C_1 to C_2 may complicate rather than amplify our understanding of the discovery process, especially the evaluative stage. DT as an independent resource makes this point clear. What is also clear, from this discussion, is that DT cannot be reduced or subsumed to IBE.

Focusing on both the selective and the evaluative stage is a strong point in the IBE approach. Like DT, IBE requires a context of comparison. On the other hand, DT is not a mechanism for setting up the selective stage, since DT deals with only one candidate hypothesis and not the alternative one. As a consequence, its interpretation of the explanatory power of possible hypothesis as a guide for its evaluation is quite different from the interpretation of IBE. As we have seen, DT's interpretation is not less efficient than that of IBE in explaining the case of Semmelweis. The apparent lack of a treatment of comparative and selective context is not because Dignaga did not know or did not worry about the selective stage, but because he treated the problems of hypothesis competition and selection within a larger discursive context of discussion and debate. Dignaga's theory of reasoning was a theory of argumentation for debate, as most of reasoning theories of India were.[h] DT is only a part of the whole theory, focusing on a technical treatment of evidential evaluation. In a series of discussions about how to distinguish a good argument from a bad one, Dignaga studied many conditions under which a better argument can be chosen. It is quite clear that a candidate hypothesis that does satisfy the triple rule is better that an alternative hypothesis that does not satisfy o does satisfy less satisfactorily. However, a satisfactory model for the selective stage requires something more than what DT can offer.

VI. Conclusion

I have argued that DT as a bottom-up approach to discovery is alternative to the top-down approaches, irreducible to the latter. This by no means implies that DT has any priority over other alternative models of discovery, but only implies that there are epistemic virtues to a pluralist attitude, according to which different models grasp different cognitive and epistemic aspects of scientific discovery, with DT being just one of these models. The main reason of adopting the pluralism is the fact that, in scientific inquiry, a observed fact in need of explanation or, in the terminology of Nickles, a scientific problem should "be viewed as a structure of constraints containing a 'gap' (the sort of gap determining the type of problem), plus a programmatic or methodological demand that the gap be filled".[4] Different scientific practices might imply different models of discovery. This point is well shown in our comparison between DT and IBE in the section.

IBE gives us a good model through which we can understand how the best explanation is chosen from a group of candidate explanations in that stage of scientific discovery in which several explanations have been generated, whereas DF focuses basically on the generative and the evaluative stage. Defenders of IBE such as Harman and Lipton maintain that IBE is a model for all non-demonstrative inferences. However, besides the problem of how to have a clear definition of a hypothesis explanatory power, as we have seen in the case Semmelweis, not all discovery process begins with all candidate explanations already given. Abduction models the inference that take the logical relation between C and A as a reason for believing that A is a hypothetical explanation of C, and this has been shown to be a very effective strategic rule for some kinds of scientific discovery. As an inferential scheme, for instance, it has enjoyed a lot of success in artificial intelligence. However, as we have seen, many aspects of scientific discovery cannot be modeled by the top-down approach. Different approaches are possible, and DT is an alternative.

[h] About the argumentation context in which Dignaga's reasoning theory embedded, see e.g., Huang[18] and Matilal.[19]

References

1. I. Niiniluoto, Defending Abduction, in *Philosophy of Science, Proceedings*, **66**, S436-S451, 1999
2. A. Aliseda, *Abductive Reasoning: Logical Investigations into Discovery and Explanation* (Springer; 2006).
3. P. Lipton, *Inference to the Best Explanation* 2nd ed. (Routledge, 2004).
4. T. Nickles, (ed.) *Scientific Discovery, Logic, and Rationality* (Reidel; 1980).
5. S. Paavola, Abduction as a Logic of Discovery: The Importance of Strategies in *Foundations of Science*, **9**, 267-283, 2004
6. J. Hintikka, *Inquiry as Inquiry: A Logic of Scientific Discovery*, (Kluwer, 1999).
7. I.M. Bochenski, *A History of Formal Logic*, (Trans. I. Thomas, University of Notre Dame Press,1961)
8. C. Oetke, Ancient Indian Logic as a Theory of Non-monotonic Reasoning, in*Journal of Indian Philosophy*, **24**, 447-539, 1996
9. C.G. Hempel, *Philosophy of Natural Science*, (Prentice-Hall, 1966)
10. S. Paavola, Hansonian and Harmanian Abduction as Models of Discovery, in *International Studies in the Philosophy of Science*, **20(1)**, 93-108, 2006
11. P. Achinstein, Inference to Scientific Laws in R.H. Stuwer (ed.) *Minnesota Studies in the Philosophy of Science*, **5**, 87-111, 1970
12. C.G. Hempel and P. Oppenheim Studies in the Logic of Explanation, in *Philosophy of Science*, **15**, 137-75, 1948
13. I. Lakatos, Changes in the Problem of Inductive Logic, in I. Lakatos (ed.) *The Problem of Inductive Logic*, North-Holland Publishing Company, 315-417, 1968
14. Y. Ben-Menahem, The Inference to the Best Explanation, in *Erkenntnis*, **33**, 319-344, 1990
15. T. Day and H. Kincaid, Putting Inference to the Best Explanation in Its Place, in *Synthese*, **98(2)**, 271-295, 1994
16. J.H. Fetzer, Propensities and Frequencies: Inference to the Best Explanation, in *Synthese*, **132(1-2)**, 27-61, 2002
17. P. Lipton, Is Explanation A Guide to Inference? A Reply to Wesley C. Salmon, in G. Hon and S. S. Rakover (eds.) *Explanation: Theoretical Approaches and Applications*, (Kluwer, 2001, 93-120)
18. X. Huang, La reconstrucción de la teoría del razonamiento de Dignaga por medio de la lógica formal in *Diánoia*, **XLVIII(50)**, 3-40 2003
19. B.K. Matilal, *The Character of Logic in India*, (State University of New York Press, 1998)
20. R.P. Hayes, *Diagnaga on the Interpretation of Signs*, (Kluwer, 1988)

THE NEW IGNORANCE

JANET A. KOURANY

University of Notre Dame
Department of Philosophy, 100 Malloy Hall,
University of Notre Dame, Notre Dame,
Indiana 46556, U.S.A.
jkourany@nd.edu
1 574 631-6864
1 574 631-0588

Keywords: Ignorance; knowledge; agnotology; philosophy of science; epistemology; Proctor; Oreskes; United States.

Although many today speak of ours as a knowledge society, ignorance seems to flourish all around us. Even in the United States, considered one of the most advanced of societies, millions question some of the most established results of science (such as evolution, global warming, and the benefits of childhood immunization) while they fail to question other results (such as many of the claims about the safety and effectiveness of drugs offered by the pharmaceutical industry). What's more, only a small fraction (less than one-sixth) of the information gathered in the U.S. each year is even made public (the rest is kept secret by industry or government), and the information that is made public often varies in content depending on the sources consulted. And the problem, many say, is getting worse.

As a result, within the last 10 years historians of science such as Robert Proctor, Londa Schiebinger, Peter Galison, and Naomi Oreskes have been promoting a new area of enquiry — Proctor calls it *agnotology*, the study of ignorance — which they suggest is of as much relevance to philosophers and scientists and others as it is to historians (see, e.g., Proctor and Schiebinger 2008). Indeed, the suggestion is that agnotology offers a new approach to the study of knowledge, an approach at least as important as its more established philosophical sisters, epistemology and philosophy of science. According to this new approach, ignorance is far more complex than previously thought. Ignorance is not just the void that precedes knowledge or the privation that results when attention focuses elsewhere. It is also — in fact, it is especially — something socially constructed: the confusion produced, for example, when special interests block access to information or even create misinformation. Thus, Proctor has written about the tobacco industry and the techniques it used in the past and continues to use to produce doubt in the public regarding the health risks of smoking (Proctor 2013). Oreskes and coauthor Erik Conway have written about the small group of politically conservative scientists who worked behind the scenes to stall public recognition of such problems as acid rain, the ozone hole, and global warming (Oreskes and Conway 2010). Galison has written about the development of the U.S. government's procedures and motivations for keeping vast quantities of information secret ("classified") and the problems that have arisen both inside and outside government as a result (see, e.g., Galison 2008 and Galison and Moss 2008). And so on. These are wonderful, path breaking historical studies that surely exemplify an expanded understanding of ignorance. But do they also exemplify a new area of enquiry, agnotology, on a par with other areas such as epistemology and philosophy of science? And if they do, what relation might the new area bear to these other areas — to epistemology, whose focus is on knowledge rather than ignorance, and to philosophy of science, whose focus is on the knowledge of scientists, not the ignorance of the public? What might agnotology contribute to philosophy, and what might philosophy contribute back? These will be my questions.

Start with the understanding of ignorance provided by philosophy. At least in modern times, philosophy has provided a very expansive conception of ignorance. Human ignorance, of course, refers to what humans fail to know, and for much of the modern philosophical tradition this ignorance has been thought to be, or at least it has been feared to be, nearly boundless. Epistemology in the past, for example, very much followed in the footsteps of Descartes, and much of it still does, but while Descartes never thought that his evil demon had gotten the better of him, his followers have certainly thought otherwise. Indeed, neither Descartes nor any of the epistemologists after him has ever been granted a clear victory over the kinds of skeptical doubts raised by Descartes, and New York University epistemologist Peter Unger, in his now classic and highly influential book *Ignorance* (1975, 2002), has argued with great cogency that no one ever will be granted that victory. According to Unger, in fact, no one knows anything at all, and hence no one ever will be justified or reasonable in believing anything at all as well. Knowledge and rationality are simply not to be had.

The story from traditional philosophy of science has been similar. There, Hume has been the hero rather than Descartes, and Hume's problem of induction and later what might be called the problem of abduction (or inference to the best explanation) have taken the place of Descartes' demon. And although scientific realists have tried frightfully hard to justify scientists' claims to know in the face of these and related challenges, once again none has finally been granted the prize. As a result, some of the most learned figures in philosophy of science, such as Karl Popper, have candidly admitted that scientific ignorance is our fate: neither theories nor even basic statements (accepted by convention) are ever justified (see, e.g., Popper 1959, 1963). And even when the learned are not quite that candid, the result seems the same. Think, for example, of Thomas Kuhn (1962), for whom the march of science is a movement away from primitive beginnings but not toward anything at all — certainly not toward any fixed end set by nature. Or think of Bas van Fraassen (1980), for whom induction but not abduction is reasonable. For him, therefore, observation statements and empirical laws are knowable, but theories are not, even though for him observation statements and empirical laws are theory-laden with those same unknowable theories.

So, prominent traditions in both epistemology and philosophy of science have emphasized the paucity of human knowledge and the vastness of human ignorance. Of course, there have always been problems with these stories. For one thing, their standards for knowledge have been unreasonably high, involving as they do what Dewey would call a misguided quest for certainty. Perhaps because of this, no one has ever really believed these stories even with all the arguments given in their defense. At least no one has ever acted as though they did. Unger, for example, has continued to give his talks and publish his papers and teach his classes even though he has claimed to know nothing at all. And Popper, I dare say, never worried over the consequences of doing the things he did even though he claimed to have no grounds for any particular expectations. And this points up the fact that there has never been an appropriate response to the traditional stories. At least these stories never called for any response, never demanded any remedy for the ignorance they described. How could they, when their arguments called into question even the bodies that would be needed for a response. No matter. The upshot is that knowledge, for epistemologists and philosophers of science, has seemed a difficult achievement, perhaps an impossible achievement, and at any rate one that needs to be explained, to be justified, while ignorance is perfectly natural.

Agnotologists seem to be offering the opposite story. According to this new story, *knowledge* is frequently what is natural, and ignorance is frequently what needs to be explained. Think, for example, of the tobacco study of Proctor, and the global warming study of Oreskes and Conway. Cigarettes are surely harmful; that is uncontroversial — in fact, has been uncontroversial for more than 50 years. They have been linked not only to lung cancer but also to heart disease and strokes and emphysema and diabetes and a host of other diseases. And they kill six million people per year, more than all the world's infectious diseases combined. Proctor calls them the deadliest artifacts in the history of human civilization. Indeed, if all smoking were to cease tomorrow, the extraordinary death rate from smoking would still remain in effect for decades. Yet, the belief continues to be widespread that smoking is safe — or at least is safer than it used to be, or at least is safe enough to be acceptable, at least for a while; that it is an "adult choice" that connotes freedom and independence and sophistication; that it is a wonderfully pleasurable and relaxing pastime; etc., And so, cigarettes continue to be the most widely used drug in the world, with six trillion cigarettes sold every year (Proctor 2013). Global warming has also been uncontroversial for years. If no measures are taken to deal with it, the rise in global surface temperature could be as high as 11 degrees Fahrenheit during this century, and the longer we wait to deal with it, the worse the outcome will be (IPCC 2007). Yet, the general public continues to doubt that global warming is occurring. According to a Yale University and George Mason University survey conducted just last month (in April, 2013), for example, only 63% of Americans believe global warming is occurring, down from 70% the previous September because of recent abnormally cold weather in some parts of the United States (Leiserowitz *et al.* 2013). And in other countries the situation is even worse. Indeed, a 2007–2008 Gallup Poll that surveyed people in 128 countries found that over a third of the world's population has never even heard of global warming (Pelham 2009). As a result, serious action has been delayed. With all the information readily available about the hazards of smoking and the dangers of global warming, what needs explaining here is the enduring ignorance of so many in the face of all this information.

The tobacco case of Proctor and the global warming case of Oreskes are shocking. And yet, agnotological cases like these are common. Consider, for example, the situation in the United States. Books and newspapers are plentiful there; radio and television, libraries, and the internet are widely available; public education is universally available, with school attendance compulsory for most children; and 41% of the U.S. population have completed at least some form of post-secondary education, the fourth highest percentage in the world (right behind Canada, Israel, and Japan) (National Center for Education Statistics 2011, Stockdale and Sauter 2012). Yet, ignorance in the United States is rampant about even the most basic things. Take nutrition, for example, whose fundamentals anyone should be able to understand and everyone must master to stay healthy. People in the United States are *inundated* with information about nutrition. There is the new, ultra-simple, omnipresent "MyPlate" symbol (replacing the slightly more complicated pyramid symbol) put out by the United States Department of Agriculture (USDA) that graphically explains the different food groups and the proportions of each one should include in a healthy diet. There are the various USDA and other nutrition and medical websites and internet hotlines that explain the fundamentals of diet and weight management, that provide tools to help people track what they eat and help them eat better, that give them personalized plans for healthier eating, and that answer specific questions people send in (see, for example, the USDA ChooseMyPlate.gov website for all of these items). There are the numerous health and diet books and magazines and articles, the information about ingredients and nutrition ratings on food packages, and the ongoing health education campaigns such as the "Face the Fats" national consumer education campaign launched by the American Heart Association in 2007 (see its website at http://www.heart.org/HEARTORG/GettingHealthy/FatsAndOils/Face-the-Fats_UCM_301588_SubHomePage.jsp) and "The Heart Truth" national awareness campaign for women about heart disease launched in 2008 by the National Heart, Lung, and Blood Institute (see information about the campaign at http://www.nhlbi.nih.gov/educational/hearttruth/about/index.htm). And, of course, there are the lessons about health and nutrition for schoolchildren. And yet, despite all of this nutrition– and health–related information, and assistance, and encouragement, two out of every three adults and one out of every three children are overweight or obese in the United States, conditions that have been linked to diabetes and heart disease and stroke and a variety of other diseases (see the Centers for Disease Control and Prevention "Overweight and Obesity" website at http://www.cdc.gov/obesity/data/adult.html). And it is this ignorance — this startling disconnect between abundant, readily available, and easily understandable information and consistently harmful nutritional choices on the part of a majority of the population — that demands explanation.

Or consider a more complicated, but equally important case of ignorance in the United States, that of racism, especially racism against African Americans (see, for what follows, the Gallup Organization 2001; Kozol 2005; CNN 2006; Wise 2006; Associated Press 2007; Jones 2008; McCall 2010; Censky 2011; Gye 2012; U.S. Department of Labor 2012, 2013). In the U.S., information is readily available in books and articles, on radio and television and the Internet, as well as in formal studies that racism is a serious problem. What kind of information are we talking about? Consider just unemployment. Historically, African Americans have had higher unemployment rates than all the other major racial and ethnic groups. The average unemployment rate for blacks in 2011, for example, was 15.8 percent, compared to 7.9 percent for whites, and 11.5 percent for Hispanics, while current rates (just issued for May, 2013) are 13.2 percent for blacks, compared to 6.7 percent for whites, and 9.0 percent for Hispanics. And when we look at some of the research that has been done over the last decade on the subject of racism, we begin to see the story behind those statistics. For example, studies have documented the deplorable conditions in low–income schools all across the nation that serve African American as well as Hispanic students. And conditions for these students are not much better even in more affluent schools. Studies show that black and Hispanic students are about half as likely as whites to be placed in advanced or honors classes even when test scores and prior performance are comparable, and they are twice as likely to be placed in remedial classes. They are also two to three times more likely to be suspended or expelled from school even though rates of serious school rule infractions do not differ among racial groups to any significant degree. Nor do conditions get better after graduation. Studies show, for example, that people with names such as Lakisha and Jamal, names associated with the African American community, are 50% less likely to be called back for a job interview than people with "white sounding" names such as Emily and Greg, even when credentials, experience, and style of communication are the same. What's more, studies show that white men with criminal records are slightly more likely to be called back for job interviews than black men without criminal records, even when the men present themselves to potential employers in an identical fashion and are equally qualified. Indeed, whites are almost twice as likely as blacks to be preferred even when they have less experience and fewer credentials. Small wonder that black families are three times more likely to live below the poverty line than white families.

Of course, to all this must be added the data concerning race-related disparities in wages and benefits, in health care and life expectancy, in housing and environmental hazards, in arrest and incarceration rates, and in public and private social services (such as treatment in stores, restaurants, and public transportation, and treatment by police and fire fighters). Taken together, these data reveal an extremely serious racism problem in America. And yet, in poll after poll a majority or near majority of white Americans deny that racism is a serious problem. For example, a 2001 Gallup poll found that only 4% of whites thought that blacks are treated badly in the United States. A 2006 CNN poll found that only 18% of whites thought that racism is a "very serious" problem. And a 2008 *USA Today*/Gallup poll found that only a bare majority of whites — 51% — thought that racism against blacks is widespread in the U.S., though that poll also found that 42% of whites thought that racism against *whites* is widespread in the U.S. So again, what requires explanation is ignorance, not knowledge — the continual denial by a majority of whites that racism needs to be attended to in a far more serious way than it has been already.

So, agnotology seeks explanations of ignorance while philosophy of science and epistemology seek explanations — justifications — of knowledge. But agnotology goes further. It also succeeds, as prominent traditions in epistemology and philosophy of science have not, in providing these explanations, or at least it suggests that the explanations can be provided. Consider a few examples:

In his new book, *Golden Holocaust* (2013), Proctor explains how in the 1950s, when the evidence for the link between smoking and lung cancer had become overwhelming, the tobacco industry opted to "fight science with science" in order to forestall regulation and protect its profits. The industry devised a set of strategies called by Proctor the "tobacco strategies" to do this, strategies that included: funding decoy research to distract from critical questions, thereby "jamming the scientific airwaves"; organizing "friendly research" for publication in popular magazines; establishing scientific front organizations; producing divergent interpretations of evidence and even misinterpretations of evidence and even suppressions of evidence; forever calling for more research and more evidence and setting standards for proof so high that nothing could ever satisfy them; and exploiting or actually producing divergent expert opinion. In addition, Proctor's earlier book, *Cancer Wars* (1995), explains how trade associations were later organized to pursue some of the same strategies to shield still other industries from regulation — trade associations such as the Polystyrene Packaging Council, the Fertilizer Institute, the Chemical Manufacturers Association, the National Dairy Council, and the American Meat Institute. And other books that have appeared since *Cancer Wars* have produced additional details regarding the methods these industries have developed to produce ignorance in the American public. For example, *Salt, Sugar, Fat: How the*

Food Giants Hooked Us (2013), by Pulitzer Prize winning New York Times reporter Michael Moss, explains the strategies the food industry (many of whose companies are owned by tobacco giants R.J. Reynolds and Philip Morris) has used to manipulate the American public into ever less healthy food choices to generate ever more profits for the industry.

For their part, Oreskes and Conway, in their recent book, *Merchants of Doubt* (2010), explain how four distinguished scientists — Fred Seitz, past president of the National Academy of Sciences and of Rockefeller University; Robert Jastrow, founding director of the Goddard Institute for Space Studies; William Nierenberg, past director of the Scripps Institution of Oceanography; and Fred Singer, first director of the National Weather Satellite Center and founder of the Science and Environment Policy Project in his home state of Virginia — Oreskes and Conway explain how these four distinguished scientists adopted the same tobacco strategies to produce doubt and confusion in the American public regarding not only global warming but also other serious problems such as acid rain and the ozone hole. In each case they worked against a strong consensus within the international scientific community by such tactics as misrepresenting, suppressing, or attacking the results of scientific studies that supported the consensus positions; lodging personal attacks against scientists whose research was central to the development of these positions; defending competing, but weaker, positions than the ones accepted; and in general acting to create the impression that controversy still surrounds issues that the international scientific community considers settled. This time the motivation, Oreskes and Conway tell us, was not so much the safeguarding of profits as an anti-regulation, market–fundamentalist set of political commitments. Still, the activities of Seitz and the others were backed by major conservative think tanks that were, in turn, backed by the U.S. fossil fuel industry, particularly ExxonMobil.

Others (such as the present author) are investigating additional parts of the agnotological picture — including, for example, the history of race-differences research in the U.S., its support from conservative foundations and think tanks such as the Pioneer Fund and the American Enterprise Institute, and the ways such research has helped to keep U.S. racism alive. Taken together, all these studies provide insight into the vastness of the science regime that is producing ignorance in the American public.

The upshot is that science has been a central player in the production of the American public's ignorance even while it has been a central player in the production of the knowledge that the public is lacking. What's more, it has been very difficult — for the public and even sometimes for the experts — to distinguish when science is producing the one and when it is producing the other. But this suggests some answers to the questions with which we began. For although agnotologists accept (indeed, *presuppose*) that science produces knowledge — the knowledge of the harms caused by smoking, for example, or the knowledge of global climate change, or the knowledge of the elements of a healthy diet — agnotologists have largely failed to explain what *constitutes* these results of science as knowledge. Oreskes and Conway, for example, have portrayed science's peer review system and the mechanisms it employs to generate scientific consensus as hallmarks of the production of genuine knowledge. At the same time, Proctor has challenged this account, emphasizing how economic and political interests have helped to shape these mechanisms — emphasizing how, for example, the tobacco industry has organized conferences, established journals, supported certain lines of research while criticizing or even suppressing other lines of research, and so on. As a result, these peer-review mechanisms are as central to the production of ignorance as they are to the production of knowledge. Ironically, Oreskes and Conway have also challenged their account of knowledge with their many examples of how such tobacco industry strategies have been put to use in a variety of other, non–tobacco–related contexts. Meanwhile, none of these individuals has offered a more adequate account of scientific knowledge, even though the distinction between such knowledge and the socially constructed ignorance on which they focus is central to their project. Of course, what constitutes knowledge, scientific or otherwise, is just what philosophy has traditionally attempted to explain. If it has so often failed in the attempt, that may be because its methods of analysis have largely lacked the constraints on philosophical imaginings that real-life, socially important questions about knowledge and ignorance provide. But these constraints are just what the questions of agnotology now provide. In short, the way to achieve success in both enterprises — in the knowledge–focused enquiries of philosophy, epistemology and philosophy of science, as well as in the ignorance–focused enquiries of agnotology — may be to combine the resources and needs of both into a new, more inclusive area of enquiry that adequately deals with knowledge *and* ignorance: episteagnotology.

References

Associated Press 2007. "Income Gap Between Blacks, Whites Expands," *NPR*, November 13. At http://www.npr.org/templates/story/story.php?storyId=16257374

Censky, Annalyn. 2011. "Black Unemployment: Highest in 27 Years, *CNN Money*, September 2. At http://money.cnn.com/2011/09/02/news/economy/black_unemployment_rate/index.htm

CNN 2006. "Poll: Most Americans See Lingering Racism — in Others," *CNN US*, December 12. At http://www.cnn.com/2006/US/12/12/racism.poll/index.html?_s=PM:US

Galison, Peter. 2008. "Removing Knowledge: The Logic of Modern Censorship." In *Agnotology: The Making and Unmaking of Ignorance*, ed. Robert Proctor and Londa Schiebinger. Stanford, Calif: Stanford University Press.

Galison, Peter and Robb Moss. 2008. *Secrecy*. Oley, PA.: Bullfrog Films.

Gallup Organization. 2001. "Black–White Relations in the United States: 2001 Update," *Gallup*, July 10. At http://www.gallup.com/poll/9901/BlackWhite-Relations-United-States-2001-Update.aspx#Exec

Gye, Hugo. 2012. "Restaurant Racism Revealed as 40% of Waiters Admit They Discriminate Against Black Customers Because They 'Don't Tip Well' ", *Daily Mail*, April 24. At http://www.dailymail.co.uk/sciencetech/article-2134200/Restaurant-racism-revealed-40-waiters-admit-discriminate-black-customers-dont-tip-well.html#ixzz1sxTSYU6R

Intergovernmental Panel on Climate Change (IPCC). 2007. "Climate Change 2007: Synthesis Report," *IPCC Fourth Assessment Report: Climate Change 2007*. At http://www.ipcc.ch/publications_and_data/ar4/syr/en/main.html

Jones, Jeffrey M. 2008. "Majority of Americans Say Racism Against Blacks Widespread." *Gallup*, August 4. At http://www.gallup.com/poll/109258/Majority-Americans-Say-Racism-Against-Blacks-Widespread.aspx

Kozol, Jonathan. 2005. *The Shame of the Nation: The Restoration of Apartheid Schooling in America*. New York: Random House.

Kuhn, Thomas. 1962. *The Structure of Scientific Revolutions*. Chicago: University of Chicago Press.

Leiserowitz, A., E. Maibach, C. Roser-Renouf, G. Feinberg, and P. Howe. 2013. *Climate Change in the American Mind: Americans' Global Warming Beliefs and Attitudes in April, 2013*. Yale University and George Mason University. New Haven, CT: Yale Project on Climate Change Communication. At http://environment.yale.edu/climate-communication/files/Climate-Beliefs-April-2013.pdf

McCall, Ken. 2010. "Well–Off Blacks Denied Loans More Than Low–Income Whites," *Springfield News-Sun*, April 10. At http://www.springfieldnewssun.com/news/news/local/well-off-blacks-denied-loans-more-than-low-incom-1/nNBwY/

Moss, Michael. 2013. *Salt, Sugar, Fat: How the Food Giants Hooked Us*. New York: Random House.

National Center for Education Statistics. 2011 "Digest of Education Statistics: 2011." At http://nces.ed.gov/programs/digest/d11/index.asp

Oreskes, Naomi and Eric M. Conway. 2010. *Merchants of Doubt: How a Handful of Scientists Obscured the Truth on Issues from Tobacco Smoke to Global Warming*. New York: Bloomsbury Press.

Pelham, Brett W. 2009. "Awareness, Opinions about Global Warming Vary Worldwide," *Gallup World*, April 22. At http://www.gallup.com/poll/117772/Awareness-Opinions-Global-Warming-Vary-Worldwide.aspx

Popper, Karl. 1959. *The Logic of Scientific Discovery*. London: Hutchinson.

———. 1963. *Conjectures and Refutations: The Growth of Scientific Knowledge*. London: Routledge.

Proctor, Robert. 1995. *Cancer Wars: How Politics Shapes What We Know and Don't Know about Cancer*. New York: Basic Books.

———. 2013. *Golden Holocaust: Origins of the Cigarette Catastrophe and the Case for Abolition*. Berkeley: University of California Press.

Proctor, Robert, and Londa L. Schiebinger. 2008. *Agnotology: The Making and Unmaking of Ignorance*. Stanford, Calif: Stanford University Press.

Stockdale, Charles B., and Michael B. Sauter. 2012. "The 10 Most Educated Countries in the World," *24/7 Wall St.*, Tuesday, January 31. At http://finance.yahoo.com/news/the-10-most-educated-countries-in-the-world.html

Unger, Peter. 1975, 2002. *Ignorance: A Case for Scepticism*. Oxford: Oxford University Press.

U.S. Department of Labor. 2012. "The African–American Labor Force in the Recovery," February 29 Report. At http://www.dol.gov/_sec/media/reports/BlackLaborForce/BlackLaborForce.pdf

———. 2013. Bureau of Labor Statistics Economic News Release, May 3. At http://www.bls.gov/news.release/empsit.nr0.htm

Van Fraassen, Bas. 1980. *The Scientific Image*. Oxford: Oxford University Press.

Wise, Tim. 2006. "What Kind of Card Is Race?" *Counterpunch*, April 24. At http://www.counterpunch.org/2006/04/24/what-kind-of-card-is-race/

MODELS, FICTION, AND FICTIONAL MODELS

CHUANG LIU

Department of Philosophy, University of Florida
Center for Philosophy of Science and Technology,
Shanxi University, China

1. Introduction

The use of models to scientifically represent and study reality is widely recognized with good reasons as indispensible for the practice of science. Because models, unlike purely verbal representations, are justifiably regarded as vehicles of representation that are not truth-apt, philosophical questions are naturally raised concerning the nature of such vehicles and how they represent (cf. Contessa 2009, 2010 & Levy 2010, 2012). Partly because of difficulties in answering such questions, which I briefly discuss below, and partly because of the temptation from other quarters, a sizable literature is generated in recent years in which mostly positive explorations have been made into the possibility that "scientific models are works of fiction" (cf. articles in Suáres 2009 & Godfrey-Smith 2009). Representations via models, according to the view, are very much like the representations in art and literature: they "make fictions" of the target systems or phenomena, which yield fictional objects, i.e. models, upon which scientists operate and learn. Idealization and other similar strategies are often taken to be the means by which models are made (or reality is made fictional). And degrees of approximation might be taken, albeit with some controversy (cf. Liu 2004), as the measure of how good the model-representation is achieved.

This essay does not strive to solve all the problems or even the most central problem about models and fiction. It picks out one small neglected aspect in the recent discussion and attempt to clarify it and figure out its implications. The idea is very simple. We often assume without much reflection that scientific model are typically made by idealizing the target systems. Typical examples include an ideal pendulum, a museum model of the solar system, etc. We then naturally extend the assumption to such models as the Crick-Watson model of DNA molecules and Rutherford's model of hydrogen atoms. But if we ask ourselves: what idealization is introduced in making these models, the answer, at least an accurate answer, is not exactly forthcoming. What features of hydrogen atoms are stripped away by Rutherford and his colleagues in coming up with the Rutherford model? In this essay I make an argument against the idea that idealization is also the means by which models of typically unobservable systems or mechanisms are made. Here "unobservable systems" are understood as a term in the practice of science, not as a term of art in philosophy of science.

2. Why Most Models in Science Are Not Fictional

In a recent article, Giere (2009) argued for "Why Scientific Models Should Not Be Regarded as Works of Fiction." If taken as a conclusion about all types of scientific models, I disagree with Giere; and yet I agree with him that many types are not or should not be so regarded. Some of Giere's arguments I cannot concur, and yet I think more reasons can be found to support his conclusion (see also, Giere 2002, 2011).

Giere stated at the outset without argument that he agrees with some in the recent literature that scientific models may be regarded as fictional in an *ontological sense*, but looking at their *function* in the practice of science, they cannot and should not be so regarded. "Theoretical models and works of fiction are, I will argue, ontologically on a par. It is their differing functions in practice that makes it inappropriate to regard scientific models as works of fiction." (Giere 2009, 249) Judging from the sort of entities Giere regards models as being — some sort of abstract entities — we can safely assume that Giere's "ontologically on a par" means that both scientific models and fictional characters are abstract entities of one sort or another.

Perhaps Giere is not making a committed claim here, namely, he is not asserting that models are ontologically fictional but functionally not. He is simply finding such an assertion tentatively admissible or not in conflict with his main conclusion. Be that as it may, I do find ample reasons for doubting that models in general are ontologically fictional. Typical examples of ontologically on a par but functioning very different items would be such as nouns and verbs, pistols and cannons, women and cats; and typical examples of those that are ontologically not on a par would be such as numbers and atoms, names and the things they name, and perhaps humans and angles (if there be any). It is not clear at all that when we describe (with implied inevitable idealization) a pendulum, we are making the same ontologically commitment as we are making when we describe a fictional object such as a character named "Sherlock Holmes."[1]

Prima Facie, the consideration of a simple pendulum can be regarded as a natural extension of our daily practice that omits non-essential details when describing a particular object while describing the person of Sherlock Holmes is the invention of

[1]And if the worry is that a "pendulum" is a type of objects while "Sherlock Holmes" is an individual and therefore they already belong to different ontological categories (and hence this is not an appropriate example for discussing the problem at issue), then one can easily substitute Holmes with some fictional types, such as "quantonian," a gooey stuff in the movie "Monsters vs. Aliens."

a fictional character that appears to be actual to those who can manage to suspend their disbelief.[2]

Arnon Levy (2010, 2012) made a persuasive argument for distinguished two different readings for acts of modeling: *de novo* modeling vs. *de re* modeling, and for defending to some extent the *de re* reading on scientific modeling. Very roughly (since I am not making any critical comments on Levy's view in this essay), A *de novo* modeling is an act that intends to make an imaginary object out of a particular system or phenomenon and treats the result as such, while a *de re* modeling is one that operates directly on the system and generates idealized or simplified statements about its relevant aspects. Even though he defended the *de re* reading of acts of modeling, Levy didn't use it to argue against the idea that scientific models are works of fiction. Literary genres such as historical novels, in which entirely fictional details are attributed to actual individuals, appear to suggest that *de re* modeling could still be regarded as fiction making. Be that as it may, what is relevant here with regard to Giere's claim that models are ontologically fictional but functionally not is that there is really no good reason to declare defeat to the fictionists on the ontological front; no good reason to think that when talking about diluted gas as ideal gas (as in Boyle's ideal gas model), we are not talking about *real, albeit very diluted, gas* and *its idealized behavior* but rather some *abstract object* and *its true and actual behavior*. In other words, there is good reason to say that "scientific models should not be regarded as works of fiction" period; not ontologically, not functionally, not in any sense.

Giere's main argument comprises three components: scientific models should be compared to works of non-fiction rather than fiction; model-making involves "[i]dealizations, abstractions, and approximations, yes. Fictions, no[;]" (Giere 2009, 254) and conceiving models as fictional helps the enemy of science, such as the creationists, which is not something a philosopher of science with good conscience should do. I shall not comment on the last argument in this essay but focus on the first two.

It is a bit of a mystery to me that invoking the fiction vs. non-fiction distinction, vague though it is and borderline cases abound notwithstanding, has not stopped the fictionists on scientific modeling cold. The comparison is so striking that unless the fictionists are actually touting a line that regards all creative representations, verbal or otherwise, as fiction-making of reality, and there is no fundamental distinction between the fictional and non-fictional depiction of events (i.e. history and historically fiction are fundamentally the same sort of writings), it is difficult for me to see how anyone could argue for fictionism for *all* acts of scientific modeling. The reason that fiction and non-fiction are fundamentally distinct categories of writing

[2]This statement is only true when we are comparing the original or near original acts of model-making and fiction-making. If we are thinking about a secondary acts of descriptively reporting on what one read in a textbook in which a pendulum is given as an example as oppose to reporting on reading the "Adventure of Sherlock Holms," then there is indeed little difference. However, we can rarely draw any evaluation of ontological commitments by looking at such secondary acts.

is clearly given by Giere: while non-fictional writings are judged, inter alia, by empirical evidence for details of what is written, it makes no sense to demand anything like that for fictional writings.[3] This is true even when we take into account the fact that fictional writings do in a sense represent reality, but descriptions of factual details, such as a particular meeting between Newton and Huygens, that are well corroborated by empirical findings, which are necessary for acceptable non-fictional writings, are entirely absent for fictional writings. And it is precisely this aspect that many types of scientific modeling possess in common with works of non-fiction.

More will be said later about why I only say here "many types" rather than "all" scientific modeling, but a brief remark in anticipation is in order. It is equally false, as shall be seen, to say that all acts of legitimate or good scientific modeling are squarely analogous to non-fiction rather than fiction. This is so even when we notice that on many occasions scientists would call with good reason a model "a piece of fiction" that fails to qualify as a legitimate or good enough model.

There are straightforwardly fictional models in science, or so I shall argue, and they are often referred to as "virtual" models. They are perfectly good models and some disciplines in science, disciplines that rely heavily on highly abstract mathematical reasoning often cannot do without them. Giere discussed in some detail a model that Teller (2009) used to illustrate an aspect about model making in science. It is a case in electrostatics, where in order to calculate the induced charge distribution on a large metal plate (modeled as of infinite size) by a unit charge above it at a distance, d, one is asked to replace in one's imagination the metal plate with a negative charge on the opposite side of, and at distance d to, the surface where the plate was. The model with the imaginary charge is clear a fictional model in its fullest and proper sense. It is a model created to have the same effect but a simpler mechanism, or it is positing a virtual (i.e. fictional) object that renders an equivalent effect and that simplifies calculations.[4]

And Giere doesn't think this model involves any work of fiction, his reason being that there are two models involved here. An actual (or concrete) system of a charge above a metal plate is first *idealized* into a model of a charge above an *infinite* model plate, and then this model is replaced by a effectively equivalent second model, and that is the model with the imaginary negative charge on the opposite side of the infinite plate. This analysis is certainly correct and yet I do not see how the second model is not a work of fiction, unless Giere's point is that the first model, which only involves idealization, is the real and the only significant model and it does not involve fiction. However, without the second entirely imaginary model, the idealization in the first model is in vain; in fact, the idealization taken to make

[3]Essentially identical reasons can be found in Walton (1990, 70–73), where scientific textbooks are simply included in the category of non-fiction. Note, "[w]orks of nonfiction do not, in general, qualify as representation in our special sense. (70)"

[4]The fictional negative charge is supposed to be where the "lines" of the electric field from the positive charge above the metal plate would "end up." By imagining its presence at that location simplifies the finding of the distribution of the "field-lines" on the surface of the metal plate.

the metal plate infinite makes sense only if one is going to introduce the imaginary negative unit charge.[5] So, the two models taken as a whole involves idealization and fiction.

There are similar cases in geometric optics, where an object is imagined to exist behind opaque surfaces in order to help the calculation of positions of real optic images. As I said, scientists who invent and teach such models have the good sense of correctly calling them "virtual" objects or models, which in our sense, the sense used in this literature that we are now engaged in, means "fictional."

One may object to the above by saying that the fictional element in such models are clearly meant for calculational or other conveniences; the real model, the model that can be taken seriously is the first model that is the result of idealization alone. This point might be reasonably made for those models or model elements that scientists call "virtual," and yet there are models or elements that can't be so easily dismissed. To begin with, the infinite metal plate that serves as the first stage before the introduction of the imaginary charge is also something physically impossible. Should such objects be thought of as works of fiction? People like Giere would regard them as in the same category as ideal gas or an ideal pendulum: idealization, yes; fiction, no. But there are more difficult cases, as I shall discuss in detail later. When phase transitions of matter, such as from solid to liquid or from liquid to gas (or transitions in the opposite direction), are modeled in physics as singularities, to merely call them a result of idealization might not be sufficient. To account for the possibility of a phase transition in any mathematically rigorous way, the system has to be considered as an infinite system and on top of that a thermodynamic limit (TL henceforth) (finite density/volume ratio) must exist; otherwise no such transition is theoretically possible in a system. Hence, the idealization and the imaginary infinite systems with TL are not just a convenient strategy for calculation but rather a condition of adequacy for the existence of the phenomena. Unlike the imaginary negative charge, which is clearly a device for an equivalent effect, the infinite systems with TL are at least not clearly such devices. A similar situation occurs in quantum measurement. Along the decoherentist approach to a solution to that problem, a similar "no-go" theorem reigns as long as the measurement device is conceived as a finite system. Only taking TL can "terminate" a quantum measurement, which ends with classical observable values. In other words, the decoherence approach, one of the strong contenders for a solution for the quantum measurement problem, demands that measurement processes be modeled in part by imaginary systems of infinite dimension or degrees of freedom.

There is a dimension to idealization and approximation that might be explored in favor of Giere's argument against fictionalism, although it is again nothing that the fictionalists cannot overcome. Although it is certainly true that in creating

[5]If the metal plate is conceived as having edges, the "field-lines" from the positive charge will bundle up at the edges, and therefore replacing the plate with the negative below its surface would no longer be a good model.

fictional characters or events idealization is frequently involved, the way in which idealization is done is often markedly different in works of fiction from modeling in science. When a model population is conceived in epidemiology for the purpose of studying an episode of influenza outbreak, the actual population is stripped of, in the researchers' mind, the inessential and irrelevant aspects and added standardized features such as "medium health" for ever member and the members are numbered to the infinite. The purpose of such a maneuver is so that the researchers can concentrate on salient properties of a population, which usually helps to simplify and so expedite the study. Such idealizations are often necessary to the extent that without them no study is possible. The creation of fictional characters, including populations of them, is usually a very different process. Because the purpose of such a creation is to generate an imaginary person or community of persons who or which would be actual if the imaginary world exists, the creator is supposed to produce concrete, albeit non-actual, particulars for the readers or viewers to consume. It is true that whatever description a writer could possibly give or whatever texture a painter could possible create, the result is always far from detailed enough for a particular; but one should not mistaken the limited description with the created character. Not being able to paint on canvas the back of a person does not mean that the artistically created person does not have a backside, and readers may not find any mentioning of Sherlock Holmes's toes in all of Sherlock Holmes stories, and yet it is a mistake to think that Sherlock Holmes therefore does not have toes. Whatever is neglected in a fictional creation is assumed if that is necessary to complete a particular. There can't be anything analogous to such notions as the "average man" and the "normal health" which frequent scientific models. So, where does idealization appear in works of fiction? The explicitly given properties to the characters are often cleaned up, idealized characteristics that the authors or artists imagined of being actually attributable. Fictional characters are often idealized particulars and scientific models are often idealized and abstract objects, and the difference between models and works of fiction lies not in the ways in which they are conceived, for both involve heavy idealization. It lies in the products of the imagination: in fiction, they are particulars, while in modeling, they are not.

3. Typically Fictional Models in Science

By now we should agree that strong arguments are and can be made against taking scientific models in general to be works of fiction; and the strongest reason, or the best argument, for that comes from the difference of purposes: model-making in science must be regarded as primarily aiming at, or at least helping as important tools towards, discovering facts and laws of nature, while works of fiction cannot be regarded as having such an aim. While works of fiction do reveal to us important aspects of the actual world, they may not be so intended or so regarded, and they do not do so in any direct empirical ways. This difference explains the presence or absence of the demand of verification of facts and the testing of conjectures, which further implies the possibility of a *de re* reading for modeling, which does not make sense for works of fiction.

It is equally obvious that, given the points made in the previous section, science does allow for models or elements of models that are purely fictional. The imaginary negative charge in Teller's example can best be seen as *ontologically* as well as *functionally* fictional. It is a prop, in the sense of an imaginary replacement of one object by another clearly non-existent object. It is also a case of make-believe in that one can pretend that instead of an infinite metal plate, a negative charge exists in its place and proceeds to calculate the effect. Very much like in the cases of fiction, especially in staged fiction as in films and plays, we proceed *as if* there is a negative charge in the supposed location or a candle behind a convex mirror. Following through this line of argument, we may very reasonably conclude that whatever ontological and functional status a fictional character has, the same is true for those virtual or imaginary objects in modeling that are used for convenience or for equivalent effects.

To add a further point for this conclusion, we may again use Levy's *de novo* vs. *de re* reading as a test. If the case can admit a *de re* reading, it is not analogous to works of fiction, and if it does not, if it *only* admits *de novo* reading, then it is very much analogous. Does it make sense to think of a particular piece of metal plate (with a positive charge hovering above it) as a negative point charge and proceed to seriously talk about its effect? If that is possible *de re*, then it is equally possible to talk about an aggressive person as a wolf *de re*, if a description of a wolf's behavior better illustrates the person's. I assume the latter is not appropriate, and therefore our test is negative: there is no way of giving a *de re* reading of Teller's case and many other similar cases. These models are just like works of fiction proper.

A caveat is in order at this point. There is a marked difference between using imaginary physical objects, idealized or otherwise, in one's model to get certain effects and what we may called "geometrizing" (or "mathematizing") a problem for a type of actual physical systems. Imagining the solar system as one of curves and points or translating a problem of motion, as Galileo did in his *Two New Sciences*, which is used as the main example by RIG Hughes to articulate his conception of modeling (Hughs 1996), is not analogous to creating works of fiction, nor has it ever been regarded as such by scientists, who routinely do such translations ever since Galileo, if not earlier. The reason for this is simply this: the resulting geometrical or algebraic *structures* are not really objects in any proper sense, concrete or abstract. It is harmless to think of such structures as tokens of models in science, but it would be harmful if we further think of them as objects or events that may have counterparts in works of fiction. The same warning should be sounded for regarding pure relational structures in formal logic, which logicians call "models" (as in Model Theory in logic), idealized objects and events. What then, one may ask, should we regard them if not at all in the category of charges, planes, ideal gases, and rational man?

In the Galileo case that Hughes takes for what exemplifies scientific modeling, the free fall of a heavy object near the earth's surface is "modeled" by a geometrical line that represents with obvious idealization the movement of the fall. Then another line is introduced that models a motion with constant speed that equals the halfway speed of the fall and is compared with the first line. The answer is easily seen and transparently displayed in this model. Given the above analysis, we can with good reason read this situation *de novo* or *de re*. If we do with the former, the motion is not translated into a geometrical line but rather into an idealized physical trajectory of the falling body. The trajectory is *de novo* because the actual trajectory with air resistance and all that is not directly amenable to geometrization. Then, a geometric study of this trajectory is carried out and Galileo's answer follows. But if we read this situation *de re*, we can dispense with the idealized trajectory and directly think of apply geometry to the falling body. And the end result is of course the same: the obtaining of the time required for the free fall as a type of uniformly accelerated motion in the science of mechanics.

Either way, geometrization or mathematization is not part of the model-building process, although it might be crucial to modeling in general. Strictly speaking, idealization is not involved nor is it a case of imagination or make-believe in mathematization. No suspension of believe is required to solve a geometrical or mathematical problems for a physical system or event, and anybody who has had the experience of learning to do problems in elementary physics, I mean those "word problems," can testify that imagination or make-believe is the farthest of what one needs in trying to become good at it. (The problems are usually already models and what one needs is to properly translate them into a geometric or algebraic problem so that inferences in the form of calculation can begin.)

I have written a critique of Hughes's theory of scientific modeling elsewhere, but the relevant point to mention here is that what Hughes takes in his DDI account as the component of "Denotation" (i.e. the first "D") actually comprises two separate components, the idealization to a model such as the trajectory of the free fall of an object and the geometricization of the model, which allows Galileo to apply geometry in the solution of the problem.

With this caveat in mind we can go on to discuss more examples of scientific models that obviously do not belong to the type of "virtual" objects. We should keep in mind while discussing these examples this question: but are they no different from the common sort of examples we frequently encounter in the literature, such as the ideal pendulum or gas, that we should not regard as analogous to works of fiction? Are they *sui generis* and if so, are they fictional in any good sense?

Harmonic oscillator as a type of model might be thought of as posing some difficulty. Now, one might think on behalf of Giere that no difficulty exists because "harmonic oscillator" names a class of different systems, such as the pendulum system and the electric circuit, each of which a model represents in a normal (i.e. nonfictional) way. This may appear to be the right diagnosis if the matter stops at that. However, when computer simulation technology becomes available, scientists can easily construct and study a "harmonic oscillator" in a simulation that is neither a pendulum nor an electric circuit nor anything that exists in reality but simply a harmonic oscillator. So, there is a concrete model system that the term refers to that does not exist in reality, nor is it an obvious idealization of something that does exist in reality. The computer-simulated harmonic oscillator is obviously neither an idealized pendulum nor a idealized electric circuit nor... There are plenty of models in science, especially in today's science, that are like the simulated (or simulatable) harmonic oscillators. They are conjured up by scientists to study a wide variety of systems whose behavior exhibit sufficient similarities to one another such that a general/generic description is given and a virtual object as a particular thing is generated by a computational device.

The second sort of models that may create problem is a grabbag of items. We couldn't help noticing that in Teller's example, an infinite metal plate is first introduced to set up the model in which the charge distribution of an actually large but finite metal plate is represented and calculated. The size is obviously an idealization but it is clearly introduced for convenience or expedience. In other words, one could just as well use a very large but finite metal plate for the calculation (there will be negligible distortion of the charge distribution if the edges are sufficiently distant). But then the task is more difficult and the result so close to the ideal case that taking the infinite plate, though not strictly necessary, makes a lot of sense. There is a type of models in science in which taking what people call the limit of infinity — or we shall call them infinity models — is *necessary*, at least necessary for the theoretical framework to which such models are called for.[6] For the Teller example, the charge distribution won't be "ill-defined" even if the plate is not imagined to be infinitely large, and yet, as already mentioned above, there are phenomena such as phrase transitions in condensed physics where without an infinity model, the target phenomenon, such the boiling of water through which the liquid turns into vapor wouldn't even "occur."

Because of its importance to the issue of "fictionalness" of scientific models, let us look at some details of this type of infinity models for phase transitions and critical phenomena (PT&CP henceforth). To some extent this type resembles the harmonic oscillator I mentioned earlier. Unlike the oscillators, this model does not have a simple equation (or set of equations) that can be instantiated to different systems with the substitutions of variables. Different systems or phenomena, such as different as gases and magnets or as different as evaporation and ferro-magnetization, have different models and equations, and yet the idea and *structural features* of this model for PT&CP applies to all those systems that undergo PT&CP, systems. The models for PT&CP represent a host of diverse systems or phenomena that share common features at a critical or transitional temperature (and/or pressure); and they are usually abstract and devoid of the particularities of the physical systems they represent, and yet simulations, computerized or otherwise, can be run on the

[6] For many excellent examples of such models, see Ruetsche 2011.

models independently of any of the particulars. What is different and distinguishes such models from other abstract models is the necessity of taking the limit of infinity in accounting for the particular phenomena.

The simplest of such models is the Ising model, which can be seen either as a group of models or as a model-scheme that can be used to get individual models (cf. Stanley 1971). We have Ising models for magnets and for fluids (e.g. the lattice gas). One may construct such a model of any dimension, however, the least complex Ising model that is useful for studying PT&CP is the two dimensional one; and here is the bare basics of the model. Imagine a large two-dimensional spatial region in which N "sites" (positions in space) are spread out at equal distance, where N is a large integer. Each site is endowed with one simple binary property, s_i, and between any two sites a coefficient, J, such that for any distribution, d, of s_i $(i \in N)$, we can define a "moment" function,

$$M_d = (1/N)\sum_{i=1}^{N} s_i(d), \tag{1}$$

and a near-neighbor "connection" function

$$C_d = \sum_{<i,j>} -Js_i(d)s_j(d), \tag{2}$$

for all the adjacent pairs $< i, j >$ of sites in the lattice. This function has the same dimension as the Hamiltonian of the system.

Imagine again a global parameter T that is associated the sites' innate ability to switch from one to the other value of s_i, such that it and J are the only two global parameters of the lattice that determine the probability of any given distribution of the values. With such a model, it is known that the Boltzmann distribution applies, and therefore, the probability of a given distribution, d, should be,

$$p_d = Z^{-1}\exp[-C_d/kT], \tag{3}$$

where k is the Boltzmann constant, and Z the partition function that provides the weight to the probability such that it adds up to unity. This model can be used to study a magnet (e.g. for its ferromagnetism) if the binary values of the sites have + or −; or we can use it to study fluids if the values are 1 or 0 (occupied or not occupied). And the model is only good for thermal systems in equilibrium *near the critical temperature*. It would make no sense to use it for anything far from that temperature regime. It is a remarkable fact that near the critical temperature systems that are otherwise radically different in physical appearance are remarkably similar or essentially the same thing, endorsing the cliché: "it's nothing but a bunch of molecules in empty space".

Here is in a nutshell what the "problem of PT&CP" is and how it is resolved in mathematical rigor; and how in this resolution we see the specialness of the "infinity models."[7] Thermodynamic quantities of a system are represented in thermo-statistical physics as the weighted (e.g. with a probability measure as in eq. (3)) averages of the system's molecular quantities. Mathematically, we can express them as various derivatives of the partition function Z for the system. (For instance, the pressure p is $p = -[\partial F/\partial V]_T$ (i.e. partial derivative under constant temperature T, where $F = -kT(\ln Z)$ is the free energy and the main connection via Z to the molecular quantities.) Phase transitions happen at a critical temperature, different for different phenomena, and are expressed as points of singularity (in terms divergent derivatives) in the thermodynamic quantities. Translating this into statistical mechanics, it turns out that phase transitions are possible only if there are positive solutions to the equation, $Z_{gran} = 0$, (which will make, e.g., dp/dV, at the critical temperature divergent).[8] However, it is proven that for any finite system of the Ising model type, no such solutions is possible (known as the "no-go" theorem of PT&CP), *unless* one imagines that the system is of infinite size. This becomes known as taking the thermodynamic limit so as to make sense of PT&CP in statistical mechanics. What is special about this and other infinity models, what distinguish them from the other typical idealized models, such as the frictionless plane and nonviscous fluids, is shown by the "no-go" theorems attached to them: *without going to infinity, one cannot even "find" the phenomena in question much less to begin studying them.*[9]

Another crucial area in physics also needs infinity models. When a quantum system in an entangled state of some observable (such as its momentum or spin) is measured for the values of that observable, the state "collapses" into a particular eigenstate of the observable in which a definite value is obtained (e.g. the spin is up or it is down). This so called collapse that destroys the coherence of a quantum entanglement, which constitutes the quantum measurement "process," is not any ordinary physical or causal process and has generated over the years competing interpretations of quantum mechanics in which the process is either explained away or made more realistic. Among the many alternative interpretations, the decoherence solution to the quantum measurement (QM) problem has, for what it worth, garnered the most support from working physicists (cf. Joos *et al.* 2003). The best version of the decoherence model of QM is to conceive the environment of any QM experiment, which according to the model makes the complete dissipation of the coherent state possible, as an infinite system. The disentanglement as expressed

---

[7]For details of how a thermodynamic function is related to a function in statistical mechanics, see any standard textbooks on the subject or see Liu 1999.

[8]This is not strictly correct, but it conveys the idea. Z_{gran} is the partition function of the grand canonical ensemble; for details see Liu 1999.

[9]A qualification is here in order. There are finite models (albeit still idealized) that exhibit PT&CP (and other phenomena that require the infinity models), for instance, the mean-field models, but interestingly, all such finite models require the introduction of some essentially non-statistical-mechanical elements in order to reproduce the phenomena in question. All exact microscopic models require going to infinity. For a different view on this issue, see Callender 2001.

mathematically is an asymptotic process that approaches its completion only as the size of the environment approaches infinity. Just as no transition in PT&CP is possible without taking the infinite limit, no definite values can emerge from QM unless a similar sort of limit of infinity is taken.[10]

These *infinity models* (models that carry "no-go" theorems for any of their finite versions) are certainly not like the virtual (or purely imaginary) models. The reason for this is simple: we may find the infinity models hard to believe because it involves something abstract and/or impossible, while on the contrary, the virtual models are all ordinary objects, e.g. the negative charge or the a lighted candle, that are placed where they are not. We pretend that they are there for some equivalent effects, and that's why they are so much like the fictional characters. The infinity models are not at all like that and therefore not analogous to fictional characters in that ordinary sense.

A more difficult challenge to meet is to answer the question: why are such models so different from the more mundane examples, such as the frictionless planes and ideal gas? The answer is actually quite simple. When we are dealing with the sort of idealization that amounts essentially to *neglecting the negligible*, such as ignoring the small effect of friction on a surface or in the air and ignoring the sizes of, and (non-collisional) interactions among, the molecules in a diluted gas, the resulting "unrealistic" models, because of their extreme simplicity, are easily modified towards more realistic models. In most cases, what needs to be done mathematically is to introduce some small parameters into the appropriate places in the equations of motion or of state such that their adjustable values may be obtained empirically (i.e. through matching with the experimental data). The prime example of this can be seen in the transition from Boyle's equation of state for ideal gas to van de Waals's equation for "real" gas:

$$pV = kT \Rightarrow (p + a/v^2)(v - b) = kT, \tag{4}$$

where the "a" and "b" are the modification factors for interaction and size, respectively. Two necessary conditions for this type of idealized models can be gleaned from these and other examples: (i) the modification factors must be introducible at the same level of description that the original idealized models are on; and (2) the factors must be introducible into the same equations we get from the original models. These are required to ensure that what is "idealized away" is indeed negligible and the resulting models of such idealizations are close to being correct in the right direction.

None of the infinity models I discussed above satisfy either of these conditions. The only alternative to the Ising model and its derivatives, as I mentioned above, is

[10]There is also a "no-go" theorem for QM cases, which is known as the von Neumann-Stone theorem and which roughly says that for quantum systems represented in a Hilbert space, no physical processes can reduce an entangled state to a mixture of eigenstates.

the mean-field type models for PT&CP. It begins with a molecular model but then an over "mean," i.e. average, field is added to mimic the intermolecular interactions that is of long-range. This field is *macroscopic* (i.e. not quantum-mechanical) and the equations that show CP below the critical temperature are not the same type of equations as Ising type models give.

Again, we can think of describing real but diluted gases as ideal gas in the *de re* mode, for by the acts of neglecting small noises or filling up small cracks or intervals, we expect that whatever properties we study of the target systems, their values make corresponding negligible differences between them and their models; and often less idealized models can be constructed if necessary by introducing perturbational parameters. This is markedly not true with the infinity-models I discussed above. The phenomena, whether it is phase transition or quantum measurement results, won't appear without taking the limit of infinity. That means at the very least that the "idealization" taken in these models are not about negligible differences of values nor is it easily removable, even in our imagination. Without taking the limit of infinity, we don't have the effects of PT&CP nor do we have classical values of measurement (from the decoherence model of quantum measurement), and not even anything close to such effects that can be imagined. This is the force of the corresponding "no-go" theorems: the desired phenomena do not appear in any finite systems. For this reason, these models fit better in the *de novo* mode of modeling.

In this sense, and this sense alone, I regard these models as of a special sort in science and think that they are closer to fictional objects than all the other models that we find often discussed as examples in the current literature.

4. Modeling the Unobservable

In this section I pick up the idea briefly mentioned in the introduction section that when scientists construct models for the "essentially" unobservable systems, they are doing something that is fundamentally different from what they do when modeling the observable.

This essay is not the place to engage in the debate between the realists and the empiricists about the division, or the lack thereof, between the observable and the unobservable. For the sake of understanding the nature and uses of scientific models, what we need is a clear concept of how scientists regard the division and how that division determines in part the types and nature of the models they build and use.

There is a sense in which any object as a thing of itself (aka Kant's "Ding-an-sich") may be regarded as unobservable (if not indeed "unknowable"), if one is thinking about the long-standing tradition in metaphysics of distinguishing Reality from Appearance. The Golden Gate Bridge between San Francisco and Marin County is not observable even in broad daylight on a clear day from a hilltop in the Presidio, if instead of its appearance we are talking about its reality or essence.

But for the engineers or research scientists who want to build a scale model for the bridge, it would not occur to them that they are actually building a model of something that they cannot see. It is true that a scale model of the Golden Gate Bridge may "misrepresent" many of its aspects, some of which are indeed not observable, because not having the same set of molecules relating to one another in the same way as those in the bridge means that many microscopic details of the bridge will not be represented in the model (even if the same kind of material is used to make it). But those are not the relevant aspects that engineers care for their scale models, and therefore, they wouldn't regard their task of modeling in this case as of building a model for something "unobservable."

Moreover, when models of this type — models for what the model-builders themselves regard as of observable systems or phenomena — are built, clearly known and articulated steps of idealization are taken to make sure that those and only those required features under scrutiny are accentuated in them. Because of the idealization involved, approximation has to be taken into consideration when results obtained from the study of a model are applied back to what it represents. This double procedure in opposite direction of idealization and approximation has been the trademark of a variety of modeling practices in science and engineering. But it is not for all such practices, as I have been arguing and shall further argue below.

When philosophers of science talks about modeling and idealization they often assume that one cannot do the former without the latter. Degrees of idealization may vary greatly from one job of modeling to another, but some amount of idealization must be involved, for otherwise there wouldn't be a point of constructing a model. However, if one seriously reflect on what and how such models as the Crick-Watson model of the DNA molecule or Rutherford's model of the hydrogen atom are created, it is not clear that specific idealizations — in its proper sense of removing or adding known elements so as to make the model simpler than its target — are indeed taken there. We can probably safely assume that the models in these and other similar cases are simpler than their target systems, but what we don't know, and should not assume that we know, is whether the simplification is gotten through idealization (cf. Lakhtakia 1996, Watson 2001, and also Karp 2013, ch. 13).

People who are not convinced that anything can be said intelligently, if intelligibly at all, about the division of the observable and the unobservable, people who tout a hardline version of scientific realism, may object by saying that there is clearly a sense in which we can observe the structure of a DNA molecule or a hydrogen atom, and it's after such observable structures that the models are made as a result of precise idealization. The objection is flawed in that the Crick-Watson model of DNA is a mechanical/combinatorial model while its target is no such system. We may be able to "observe" the combinatorial patterns of the bases in a DNA molecule, but how the bases are chemically or even biologically connected and what makes such connections work is not represented in the model. There is no sense in which neglecting these aspects in the model is the result of idealization, anymore than my daughter's use of a piece of Paz candy dispenser as his dad in a play that very much simplifies the interactions of the family members may be counted as an act of idealization. It is fairly obvious that when we have to consider such practices of modeling in the sciences we couldn't and shouldn't simply agree with Giere, as I quoted him earlier, and say that the results are, "Idealizations, abstractions, and approximations, yes. Fictions, no." (Giere 2009, 11)

The point should be more obvious for Rutherford's model of hydrogen atoms. The "solar-system" model that Rutherford advocated for sometime in the earlier 20th century can hardly be viewed as a model of idealization or abstraction. Sure, we have some experiments regarding the structure of atoms from Rutherford and his research group; when the alpha particles from high-speed bombardments are scattered afterwards in a certain pattern, the interior of the atoms that created such a pattern is conjectured with good reason of the sort that is similar to a "solar-system", and hence such a model for the interior of the hydrogen atom. A piece of brilliant analogical reasoning, perhaps but a carefully constructed piece of idealization or abstraction, it is not.

One may object at this point by asking: what is the difference between analogy in modeling and idealization or abstraction? Are they more or less the same sort of practice? No, they are definitely not the same sort of method (in the broad category of scientific methods) and the results, e.g. models, we obtain by using them may not be of the same nature. To see this point, let us begin with Rutherford's solar-system model for hydrogen atoms. In a rough and ready manner, let us reflect for a moment on how that model might have been conceived.[11] There was evidence for the existence of electrons as small particles that can be freed from atoms by bombardment; and since atoms are neutral in charge, there must be positively charged elements in them. Rutherford's scattering experiments seemed to show that most of the space in an atom is "empty," which implies that the positive charge must also be concentrated on small particles as well. The nature of electric charge, namely, how opposite charges attract and same charges repel, was also well known. Given the size and the empty space of an atom, the electrons must be moving rapidly around the nucleus of positive charges in order to maintain that space. From these pieces of evidence and perhaps the unconscious awareness that the same patterns are often repeated at different scales in Nature, Rutherford's model for atoms was born. It is obvious that nowhere in this process is idealization involved. Surely, the electrons might be idealized as point particles and there are some other such idealization taken into consideration, but none of them are essential

[11] The argument here is necessarily speculative for two main reasons. First, it is very unlikely that Rutherford or any of his collaborators would have left any remarks on whether it was "idealization" or "analogy" or any other method that he actually used in coming up with the model. Second, that is probably because he and his collaborators didn't even think about such a question when they were constructing the model. They might be aware that some sort of analogous reasoning was involved, but what sort exactly and whether is might be regarded as a species of idealization couldn't possibly be on their mind at any time of their career.

to the construction of Rutherford's model. Rather, it is a brilliant piece of Analogical Reasoning. This is modeling in the conception of Mary Hesse, as in her book *Models and Analogies in Science* (1966).

According to Hesse, the main character of analogical modeling is setting up a system that resembles something we are already familiar, or at least know better, and that share properties with the modeled system that appear to be essential to it. That was how electromagnetic waves in aether are modeled as waves in liquid in the 19th century physics (cf. Hesse 1966, 11). Nobody could possibly tell what waves in aether look like because nobody could imagining "finding out" by some theory independent method what aether is like, in the way in which one could at least imagine that for liquid or air. Without even being able to imagine what aether waves are like, it does not really make any sense to say that one could come up with a model for them via idealization, meaning taking away, or adding, aspects of the original that are in principle negligible. What small bumps must be taken away from the aether or small gaps or holes be filled or viscosity among parts removed, what any of these could be carried out in order to come up with an aether model for electromagnetic waves?

The same can be said and must be true of equal measure regarding the modeling of Rutherford's atoms. Rutherford and his colleagues did get glimpses of what the interior of atoms is like from various experimental results, including, most crucially, his own, and yet his model can only be viewed as the result of analogical reasoning. And it had to be so primarily because it was not feasible to come up with a model by, among other things, acts of idealization.

Again, this is radically different from how models for pendulums or the solar system are made. With Boyle's model of ideal gas, different options are possible, because it might well be a borderline case. It all depends on whether it is plausible to regard the idea of diluted gases as comprising mostly free molecules as deriving from "observation" or from speculation by means of analogy. I regard the former as most reasonable and therefore with good reason we should take the "ideal gas" model for what its name signifies: a product of idealization. The mechanical differences among difference molecules are neglected, their sizes are also neglected, any interactions among the molecules other than collision are also neglected, and so on. Voila, out drops the model, a perfect example of a product of idealization.

Even though there might not be any philosophically clean way of "proving" the point, namely, while the interior of a tank of diluted gas is regarded as observable, the interior of an atom is not, it should be quite obvious that the model for the former is a result of idealization while the latter is not; and furthermore, modeling by idealization and modeling by analogy are two fundamentally different methods in science.

Lest I be misunderstood in saying that only in the models for the unobservable analogies are used, the following needs to be said. Even though modeling the unobservable essentially involves analogical reasoning in one aspect or another, it does not imply that analogical reasoning cannot be involved in modeling the observable.

Modeling traffics on freeway by a hydrodynamic system is not unheard of nor necessarily unjustified; modeling the blood flow in animal arteries with the same type of hydraulic system may also be useful and justified. However, there is often a difference. For the observable systems, modeling with analogy is usually accompanied with idealization as well. And often the two processes are intertwined, so to speak. Modeling by analogy the blood flow in an animal's artery by water coursing through rigid pipes might be justified if it is a good idea, with independent justification, to neglect the flexibility of the artery, among other things. Otherwise, the analogy fails and the model may be judged as inadequate. This is not possible for the unobservable. It is not possible to make any judgment in the case of Rutherford's model for atoms, for example, on whether the elliptical orbits of the electrons revolving around the nucleus, which is obtained by an analogy with the solar system, is the result of neglecting some other actual subtle forces or influences inside an atom.

Hence, when it comes down to the question of what is so special about the conventional "unobservable" and its modeling, it has to be the fact that acts of idealization, whose use is in fact widespread and thought of by philosophers as necessary in all cases of scientific modeling, are and must be essentially absent.

5. Fictional Models for the Unobservable?

We now return to our central question concerning types of fictional models in science and engineering. Should we add the models for the unobservable to the list of models that I argued are definitely fictional in Section 3?

Models for the unobservable, even if fictional, are not fictional in the same sense as the "virtual" models discussed before. Those models are fictional not because they are novel but because they are by stipulation not real; the negative charge that is imagined to exist beneath the metal plate is no different from any other negative charge. This is typically not so for the models that represent unobservable systems.

They also appear to be different from the infinity-models: for one thing at least, they are in general by no means physically impossible. If it were not for the generic feature of being models for the unobservable, they are just like regular models. Imagine that there are intelligent beings that can change their sizes at will and therefore could on demand shrink themselves to the same size as electrons and get into an atom and "observe" what it is like in there. Afterwards, they grow back to our sizes and show us either by verbal or pictorial depiction what the interior of an atom is like. From that we can construct a model to represent an atom. The model as a result of such a process would indeed be an ordinary model from idealization and has nothing to do with fiction, as Giere rightly pointed out. The same thing could be imagined for any one of the models we discussed in the previous section, and models for the unobservable would not be ones from analogy but from standard idealization procedures.[12] In other words, models for the unobservable are, otherwise, regular models.[13]

The "virtual" models and the infinity-models can be regarded as properly fictional because of the following extra layer of reason. There is in general no sense in which these models are subject to empirical testing for their existence. They are not really made to produce testable results such that experiments can be arranged to check them and, positive or negative, the results can then be used to support or weaken the existence claims for the postulated objects. In the case of virtual models, what can and often need to be tested are the hypotheses or models before the virtual ones are put in place (as equivalent alternatives). This is certainly the case for the virtual negative charge in Teller's example; and if we think of such devices in the science of mechanics as "virtual displacement" or "virtual path" of motion, the same point applies. The devices are almost by default used exclusively in the part of modeling where deductive demonstrations are called for; for example, in R.I.G. Hughes's DDI account of modeling we discussed earlier, the second D is for "Demonstration" and is where virtual models may appear. The same can be said to the infinity-models. There is no sense in which experimental results can be used to confirm or reject a hypothesis of PT&CP as singularities. That hypothesis is already given in thermodynamics, where taking thermo-systems as continuous systems, phase transitions are all described as singularities in the otherwise continuously differentiable isotherms. The infinity-models are put in for the sake of "recovering" from statistical mechanics of atomistic systems the thermodynamic results that are well confirmed by experiments. In this sense, the infinity-models are also devices for demonstrations, very much like the virtual models.

The same *cannot* be said to models for the unobservable, at least not in general, not for all such models. While it is not clear whether one can really experimentally test an aether model, it is certainly expected that the type of models to which Rutherford's model for atoms belongs must be subject to experimental testing. And so is the Crick-Watson model for DNA molecules and the like.

But, to bring back Giere's original questions again, are models for the unobservable works of fiction? We've said that those "virtual" models should be regarded as fully fictional, partly because they do qualify as products of imagination or make-believe and partly because the scientists who use them regard them call them with

[12]For the aether model of electromagnetic waves, the imagined beings need to be more fanciful than most science fiction writings. The creature would have to be able to change its characters from normal material made of atoms, say, to extraordinary material such that it could perceive the medium in the way we perceive regular material. And then it has to be able to change back to the atom-based being like us to tell its story. It is a big question as to whether this is physically possible.

[13]See van Fraassen 1984 for one of early defense for a distinction between the observable and the unobservable. And the foregoing discussion assumes that "what it is like," namely, phenomenological experience, for humans is categorically different from, say, computer-generated data-streams. So we assume that building a small enough mechanical or biological recording device and send it into an atom wouldn't do the trick. The backdrop for such an assumption partly derives from Russell and partly from non-physicalist theories of mind.

good reason "fictional." However, they are actually very different from the usual fictional characters we enter in works of fiction. In almost no occasions, we would say, a legitimate fictional character is created as an ordinary non-existent individual whose creation is for the purpose of accounting for some equivalent social phenomenon that is known to have taken place without it. Hence, such models are fictional without being analogous to fictional characters or objects in literary fiction.

Models for the unobservable are different; they are certainly not by default there to represent non-existent systems or phenomena for some practical purposes of theorizing, and yet it is not clear, at least from what I have argued in the previous section, that what they represent are non-fictional. From certain perspectives on the history of science, we see many past models in science, which are now abandoned, that might be interpreted as fictional (from a hindsight), models such as the phelogistic model for combustion, the caloric model for heat, and the aether model for the optic medium. Their existence certainly had a lot to do with the unobservability of the systems they were used to represent, at least for the time when they were held, and so there doesn't seem to good reason to say that our current models for similar unobservable systems won't be of the same nature, if there is something in common among all these types of models throughout the ages.

If they are regarded as fictional, then they are analogous, unlike the "virtual" models, to fictional characters in works of fiction. Take the phelogistic model of combustion for example; there is no phelogiston as constructed by the theoretical model, just as there is no Madam Bovary as an actual individual, and yet the model did represent the reality of combustion in the sense that by figuring out how the substance phelogiston behave among itself and towards other objects and substances, we understand to some extent why and how combustion occurs in nature. Similarly, by reading Flaubert's *Madam Bovary*, and figuring out how the character Bovary behave in her environment, we understand to some extent why and how 19th century French provincial folks acted the way they did. No doubt that there is a huge surface difference between scientists who postulate and work with their models and writers create their characters, for the former *intend* for their models to represent real stuff while the latter *intend* for no such a thing. But that does not automatically determine the ontological sameness or differences of the results of their creation. If the foregoing arguments stand, what scientists have always believe to be one thing, models representing actual systems that are like what they are said to be like, is actually something else, something that do represent reality but only represent it in the way fictional characters represent the slice of society that they are created to represent.

To show that some writers do believe that their works represent reality, here are two quotes from two novelists.

Ah! sachez-le: ce drame n'est ni une fiction, ni un roman. *All is true*, il est si véritable, que chacun peut en reconnaître les éléments chez soi, dans son coeur peut-être.

— Honoré de Balzac, *Pére Goriot*

That's what fiction is about, isn't it, the selective transforming of reality? The twisting of it to bring out its essence?

— Yann Martel, *Life of Pi*

References

Callender, C. (2001). "Taking Thermodynamics Too Seriously," *Studies in History and Philosophy of Science Part B* **32**:539–553.

Contessa, G. (2009). Scientific Models and Representation. Steven French and Juha Saatsi (eds.) *The Continuum Companion to the Philosophy of Science.*

Contessa, G. (2010). Scientific models and fictional objects. *Synthese*, 172:215–229.

Frigg, R. & J. Reiss (2009). The philosophy of simulation: Hot new issues or the same old stew? *Synthese*, 169:593–613.

Frigg, R. (2010). "Models and Fiction," *Synthese*, **172**:251–268.

Frigg, R. (2011a). Fiction in Science, in Woods, J. ed. *Fictions and Models: New Essays.* Munich: Philosophia Verlag (forthcoming).

Frigg, R. (2011b). "Fiction and Scientific Representation," (preprint).

Giere, R. (2009). Why Scientific Models Should not be Regarded as Works of Fiction. *In Fictions in Science: Philosophical Essays on Modeling and Idealization.* ed. Mauricio Suárez, 248–258 (Routledge).

Giere, R. (2002). How Models Are Used to Represent Reality. *Philosophy of Science*, S742–S752.

Giere, R. (2011). Representing with Physical Models. In Paul Humphreys & Cyrille Imbert (eds.), *Models, Simulations, and Representations* (Routledge).

Godfrey-Smith, P. (2006). The Strategy of Model-Based Science. *Philosophy of Biology*, 21:725–740.

Godfrey-Smith, P. (2009), Models and Fiction in Science, *Philosophical Studies*, 134: 101–116.

Karp, G. (2013). *Cell and Molecular Biology: Concepts and Experiments*, 7ed., Wiley and Sons.

Lakhtakia, A. (ed.) (1996). *Models and Modelers of Hydrogen.* World Scientific.

Levy, A. (2012). Models, Fictions, and Realism: Two Packages. *Philosophy of Science* 79: 738–748.

Levy, A. (2010). Fictional Models De Novo and De Re, (manuscript).

Liu, C. (1999), "Explaining the Emergence of Cooperative Phenomena," *Philosophy of Science*, **66**:S92–S106.

Liu, C. (2004). "Approximations, Idealizations, and Models in Statistical Mechanics," *Erkenntnis* **60**:235–263.

Ruetsche, L. (2011). *Interpretating Quantum Theories*, Oxford University Press.

Stanley, H. E. (1971), *Introduction to Phase Transitions and Critical Phenomena.* Oxford: Clarendon Press.

Suáres, M. (2004). "An Inferential Conception of Scientific Representation," *Philosophy of Science.* **71**:767–779.

Suáres, M. (2009a). Fictions in Scientific Practice, in *Fictions in Science: Philosophical Essays on Modeling and Idealisation*, Routledge, 1–15.

Suáres, M. (2009b). Scientific Fictions as Rules of Inference, in M. Suárez (ed.), *Fictions in Science: Philosophical Essays on Modeling and Idealization*, Routledge, 158–178.

Teller, P. (2001). Twilight of the Perfect Model Model. *Erkenntnis.* 55:393–415.

Teller, P. (2009). Fictions, Fictionalization, and Truth in Science, in Mauricio Suárez (ed.),

Fictions in Science: Philosophical Essays on Modeling and Idealization, Routledge, 253–247.

Toon, A. (2010). The Ontology of Theoretical Modelling: Models as Make-Believe. *Synthese*, 172(2):301–315.

Toon, A. (2010). Models as Make-Believe. In R. Frigg & M. Hunter (Eds.), *Beyond Mimesis and Convention, Boston Studies in the Philosophy of Science*. 71–96. Dordrecht: Springer.

van Fraassen, B. (1984). *The Scientific Image*, The Clarendon Press, Oxford.

Watson, J. D. (2001). *The Double Helix: A Personal Account of the Discovery of the Structure of DNA*, Touchstone.

A CONTEXTUALIST INTERPRETATION OF MATHEMATICS

JIE LIU*

Research Center for Philosophy of Science and Technology
Shanxi University, Taiyuan, P.R. China

The nature of mathematics has been the subject of heated debate among mathematicians and philosophers throughout the ages. The realist and anti-realist positions have had longstanding debate over this problem, but some of the most important recent development has focused on the interpretations; each of the above positions has its own interpretation of the nature of mathematics. I argue in this paper a contextualist interpretation of mathematics, it elucidates the essential features of mathematical context. That is, being integral and having concrete structure, mathematical context is a recontextualizational process with determinate boundary.

Keywords: Mathematical context; integrity; boundary; recontextualization.

Introduction

Along with the influence of the philosophical debate between realism and anti-realism, recent research on mathematics has also concentrated on the interpretation of these two positions. Each has its own interpretation of the nature of mathematics.

As known, mathematical realists or Platonists believe that "mathematical entities like number, set, and function exist independent of the human mind." Without a doubt, such a position can strengthen our belief in pursuit of mathematical truths and its direct application in scientific research. However, there are some serious challenges that realists have to face, which have been clarified in Benacerraf's two distinguished articles. 'What numbers could not be' and 'mathematical truth'. In 'what numbers could not be', He argues against a set theoretic Platonism on the ground that what is important about numbers is the abstract structures they represent rather than the objects that number words ostensibly refer to. In 'mathematical truth', Benacerraf formulated a much more general epistemological challenge to Platonism. Platonist insist the existence of abstract mathematical objects, but such objects are epistemologically inaccessible because they are causally inert and beyond the reach of sense perceptions. This challenge is almost a devastating blow to realism. To address this challenge, a number of new realist theories have been

*Jie Liu was supported by National Social Science Funds in China: [11CZX022]; The MOE Project of Key Research Institute of Humanities and Social Sciences in Universities: [11JJD720011]; Research Project supported by Shanxi Scholarship Council of China [2012-028]; PSSR[2013301].

proposed. To wit: Quine-Putnam's indispensability argument, Maddy's set-theortic realism, ante rem structuralism, Neo-Fregean realism, and so forth. However, these strategies just raise new problems. As most of them are familiar stories. I am not going to discuss them in detail.

This challenge makes anti-realism increasingly attractive in contemporary philosophy of mathematics. The formalist approach (Geoffrey Hellman 2003a; Hartry Field 1980b, 1989b) is one of the most influential. The original intention of the formalist approach is to avoid the epistemological challenge. Its adherents take mathematical truths to be inferences from axiomatic systems. Reviewing Benacerraf's mathematical truth, he claims that: "as an over-all view, it is unsatisfactory — not so much because we lack a seemingly satisfactory account of mathematical truth or because we lack a seemingly satisfactory account of mathematical knowledge — as because we lack any account that satisfactorily brings the two together." This means that we should provide a uniform truth theory for mathematics and science, which should not only be a reasonable explanation for mathematics, but also should be coordinate with our understanding of the whole world. So my researches are mainly on how to provide an over-all view for both mathematics and science, which relies on the common progress of mathematics and science. In my opinion, Contextualism is one choice of this kind.

According to contextualism, to understand mathematics or science is to understand mathematical contexts or scientific contexts. The study of mathematical objects or scientific objects is the study of mathematical objects or scientific objects in context. The nature of mathematics or science can only be clarified by the essential features of mathematical contexts or scientific contexts. Generally, contexts have the following features: 1) Each mathematical context is made up of three factors: syntax, semantics and pragmatics; 2) Context is the unity of their interactions; 3) every context has a boundary; 4) Every context is continually re-contextualized.

1. The Structure of Mathematical Contexts

Any mathematical context is not a simple, isolated concept, but rather an overall system with complex structure. Specifically, each mathematical context is made up of three factors: syntax, semantics and pragmatics. Syntax is the study of symbols and their syntactic relations; semantics is the study of meanings of symbols; pragmatics is the study of the connections between the research subject, symbols and their meanings. Context is the unity of their interactions.

For example, David Hilbert set up an axiomatic system in *Foundations of Geometry*, which has often been taken as a model for mathematical practice. Its basic idea is to establish initial symbols, concept formation rules, axioms and inference rules, combine these elements to form a formal system, and then draw mathematical theorems through deductive reasoning in the system. All of the processes are related to context. First, the presupposition about initial concepts and inference rules is a result of pragmatic commitments in context, by which pragmatic commitments the meanings of mathematical objects are represented by specific mathematical symbols; second, the symbolic mathematical objects constitute well-formed formulas in accordance with the formation rules; certain well-formed formulas independent of each other are taken as the axioms in this context, from which new theorems can be obtained by inference rules. (Those inference rules in the context are also adopted by convention.) After the symbolic language of mathematics itself, the whole process involves only mathematical symbols and logical connections between symbols, which constitute the mathematical formal system. Considering only the logical relationship in a formal system is in essence the syntax of mathematics; finally, in order to reveal the essential features of mathematical objects, one must look beyond the syntactic features of mathematics to grasp the semantics of mathematical symbols, which is associated with the context. Specifically, mathematical symbols, mathematical formulas, and the deductions between mathematical formulas turn out to be derivations connecting propositions with a semantic interpretation. On the one hand, different branches of mathematics are formed by various kinds of mathematical structures through giving mathematical formal systems having semantic interpretations; on the other hand, all kinds of scientific laws can be interpreted directly by combining a mathematical formal system with the context of experience. The situation can be represented by the following diagram (Fig. 3):

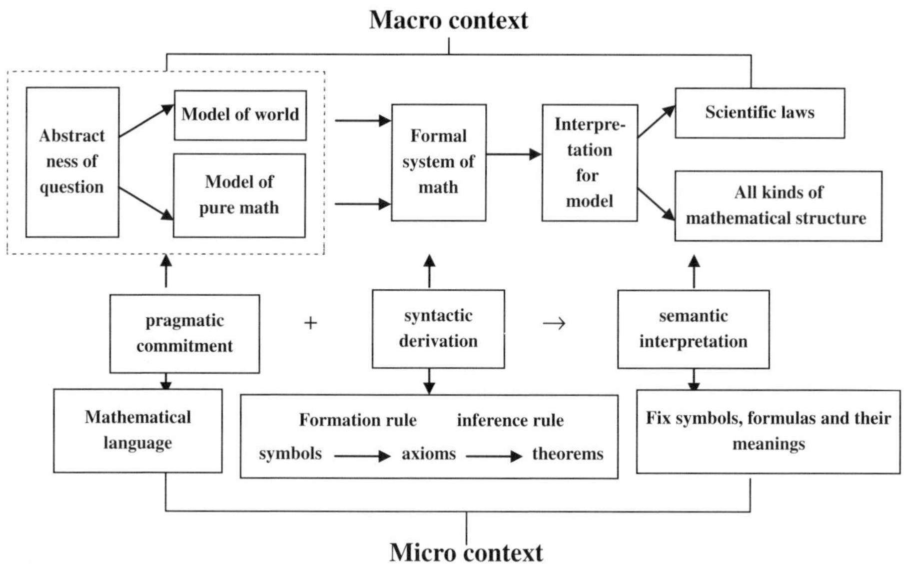

Fig. 3. Mathematical context diagram.

2. The Integrity of Mathematical Contexts

Context is the unity of syntactic, semantic and pragmatic factors. All of these are relevant to context. It can not be a context without the interaction and interrelation among these three. The representations of mathematical form are established for mathematicians' pragmatic purposes and the realization of the semantic interpretation is dependent on pragmatic purposes and provisions.

Firstly, the representations and derivations in a mathematical formal system are relevant to the pragmatic commitment in a context. The formal deductions in mathematics include premises and inference rules. As is known, all mathematical formal systems follow similar inference rules. Let Γ be a set composed of formulas of a mathematical formal system S, A is a theorem of Γ if and only if A is entailed logically by Γ, $\Gamma \vdash A$. Therefore, the relevance of the representations and derivations of a mathematical formal system to a context is embodied in the relevance of its premises to the context. For example, proposition:

At most one line can be drawn through any point not on a given line parallel to the given line in a plane.

can be taken as an axiom in Euclidean geometry context for it is equivalent to Euclid's parallel postulate. However, if we just replace the parallel postulate with the proposition:

There are many distinct lines through a particular point that will not intersect with another given line.

or with proposition:

There are no lines that will not intersect.

while maintaining the other postulates unchanged in Euclidean geometry, then a mathematical context will change into the Lobachevskian geometry context or Riemannian geometry context. Therefore, different mathematical contexts are dependent on different premises, which are relevant to their corresponding contexts. In this sense, the derivations of a mathematical formal system are relevant to its context.

Secondly, the semantic composition of a mathematical proposition is also relevant to context, and this is the most critical feature in the structure of context. In a sense, semantic composition is the key characteristic of a systematization of representational language. Above all, the spatio-temporal sequences of propositions in syntax, the concrete meaning of propositions in pragmatics, and their relations must be linked through semantic compositions. In any specific mathematical context, mathematical semantic composition depends on the pragmatic commitment in the context and subject's commitment to the formal decidability of mathematical propositions in the context. That is, the semantic interpretation of mathematics is determined by pragmatic purpose as well as certain syntactic representations in certain contexts. Considering for example the following symbolic sequence:

$$x_1, x_2, \ldots, a, f_1, f_2, =, (,), \rightarrow, \Box.$$

In a group formal system, the object constant a is the unit element, the function f_1 is inversion and the function f_2 is multiplication; however, in an arithmetic formal system, the object constant a is 0, the function f_1 is successor and the function f_2 is addition etc. Obviously, the object, constant, and functions in the above two systems have different semantic interpretations. As for the following formula $P(x_1)(x_2)(A(x_1, x_2) \rightarrow A(x_2, x_1))$, it is interpreted as 'if $x_1 = x_2$, then $x_2 = x_1$, for any natural number x_1 and x_2' in arithmetic formal system, in which the predicate A is interpreted as "="; while, it is interpreted as 'for any element x_1, x_2 in set A, if $x_1x_2 = e$ then $x_2x_1 = e$', in the group formal system the predicate A is interpreted as 'the relation between x_1 and x_2 is reversible' . It is worth noting the pragmatic purpose in a context is not arbitrary but one which is consistent with true beliefs from a background context. That means the semantic composition is in essence relative to where the pragmatic purpose comes from. Specifically, the pragmatic purpose comes from the subject's theoretical background, the cultural background and the historical background, which is embodied by the psychological state of the subject in a given context. The psychological state of the subject in given context is constituted by psychological expressions and psychological operations. The proposition that express the pragmatic purpose is realized by the agreement between psychological operation and psychological 'expressive language' as shown in the following Fig. 4:

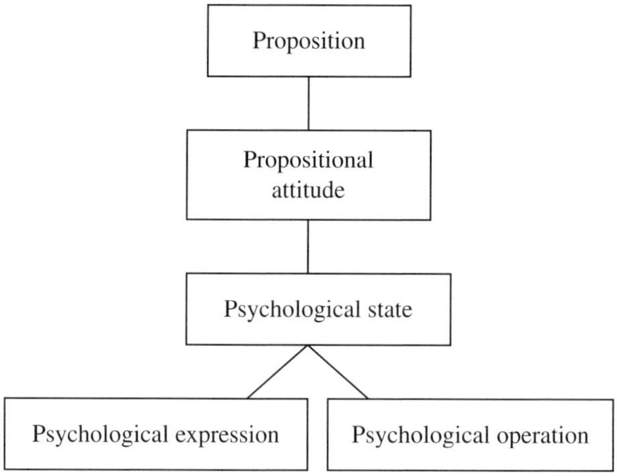

Fig. 4. Psychological structures.

The semantic interpretation depends on pragmatic commitment in context, which connects mathematical syntax with its semantic interpretation. A mathematical proposition only has meaning in virtue of the interrelation and interaction between the elements in a context. Therefore, every context is the unity of the syntax, semantics and pragmatics in it. It can be represented as following:

In short, as the unity of syntax, semantics and pragmatics, context is a kind of certain form of relation. It includes all background factors such as mathematical, theoretical, social and historical factors. This means that a mathematical object is

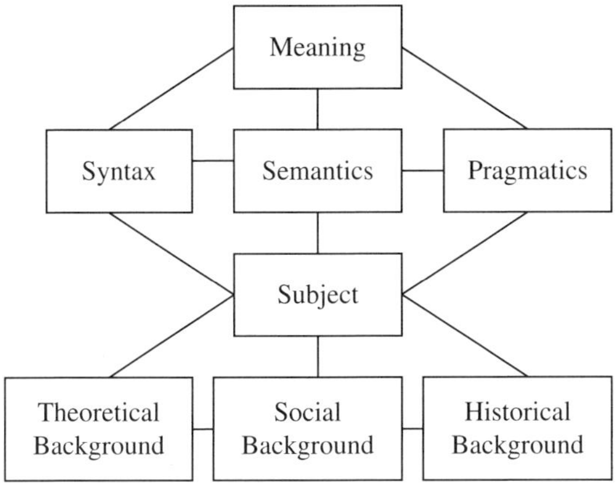

Fig. 5. Context structure.

not an individual in isolation, but one embodied in a integrated relation between object and subject in context.

3. A Mathematical Context Has a Boundary

As the platform for understanding mathematics, context has boundaries. The boundary of a context is determined by a domain of questions. Specifically, if the question is about a sentence, then the boundary of this context is this sentence; if the question is about a particular problem, then the boundary of the context is this problem; if it is a particular history, then the boundary of context is this historical process. If the boundary of semantics is exceeded, then the corresponding context will change.

In mathematics, all problems are under given boundary conditions. Once the conditions are given, then the context for solving related problems will be given. In this sense, the solutions under given conditions are the meanings of the corresponding problems in context.

Considering the following dynamic equations on time scales:

$$y^{\Delta}(t) = f(t, y(t)), \quad t \in \mathrm{T} \quad *$$

T is a non-empty closed subset of R, f is a continuous function, $y^{\Delta}(t)$ is the Δ derivative of f at t. The definition of the Δ derivative runs as follows:

For a given function $f : \mathrm{T} \to \mathrm{R}$, $t \in \mathrm{T}$, if there exists an $\alpha \in \mathrm{R}$ such that for any $\varepsilon > 0$ there exists a neighborhood U of t satisfying

$$|f(\sigma(t) - f(s) - \alpha(\sigma(t) - s)| \leq \varepsilon|\sigma(t) - s|, \quad \text{for all } s \in U,$$

then we say α is the Δ derivative of f at t, denoted $f^{\Delta}(t)$.

In different mathematical contexts, the boundary of the context will be determined by the meanings of T. For equation *, if T = R, then * reduces to a differential equation:

$$y'(t) = f(t, y(t)),$$

which describes a continuous process;

if T = Z, then * reduces to a difference equation:

$$y(t+1) - y(t) = f(t, y(t)),$$

which describes a discrete process;

if $T = [0, 1] \cup N$, then * reduces to dynamic equations on time scales

$$y^{\Delta}(t) = f(t, y(t)),$$

which describes a process with both continuous and discrete aspects.

In fact, it is the constant changes of the boundary of context that leads to the development of mathematics — recontextualization of mathematics.

4. The Recontextualization of Mathematics

As Richard Rorty said,

> *"At a certain point in this process it becomes useful to speak of 'recontextualization.' The more widespread the changes, the more use we have for the notion of 'a new context.' This new context can be a new explanatory theory, a new comparison class, a new descriptive vocabulary, a new private or political purpose... the possibilities are endless."*[1]

Mathematical context is not static but dynamic. The larger the changes in depth and breadth of context, the more profound and more diverse meanings the new context will have. The movement, change, and development of context are the process of 'recontextualization'. Mathematical context is continually re-contextualized, and changes continuously.

However, contextualism is different from Kuhn's paradigm theory, which emphasizes that scientific revolutions and scientific progress involve incommensurable paradigm shifts. Contrasted with Kuhn's paradigms, different mathematical contexts are related to each other. In this sense, the development of mathematical theory involves not only continuous accumulation but also the composition of the revolution. In order to clarify this, let us consider the following example on the integrability condition of functions.

G. Riemann discussed the problem of integrable functions when he was studying trigonometric series, and proposed the following definition of integral[2]:

Assume that $f(x)$ is bounded on $[a, b]$. Insert arbitrary $(n-1)$ points in $[a, b]$,

$$a = x_0 < x_1 < x_2 < \cdots < x_{n-1} < x_n = b,$$

[1] Rorty, R. Objectivity, Relativism and Truth. Cambridge: Cambridge University Press, 1991: 94.

[2] Zhou, Minqiang. Real Function Theory. Beijing: Peking University Press, 1998: 7–8.

which divide $[a, b]$ into n intervals

$$[x_0, x_1], [x_1, x_2], \ldots, [x_{n-1}, x_n].$$

The length of the intervals respectively is

$$\Delta x_1 = x_1 - x_0, \quad \Delta x_2 = x_2 - x_1, \ldots, \Delta x_n = x_n - x_{n-1}.$$

Let $\xi_i(x_{i-1} \leq \xi_i \leq x_i)$ be an arbitrary point in $[x_{i-1}, x_i]$, and

$$S = \sum_{i=1}^{n} f(\xi_i)\Delta x_i,$$

denoted $\lambda = \max|\Delta x_1, \Delta x_2, \ldots, \Delta x_n|$. if $S \to I$ as $\lambda \to 0$, where I is a constant which is independent on the division of $[a, b]$ and ξ_i in $[x_{i-1}, x_i]$, then I is a definite integral of $f(x)$ on $[a, b]$, noted as $\int_a^b f(x)dx$, i.e.

$$\int_a^b f(x)dx = I = \lim_{\lambda \to 0} \sum_{i=1}^{n} f(\xi_i)\Delta x_i.$$

From the above definition, we can see that the non-continuous point of integrable function must be surrounded by intervals, the length sum of which is arbitrarily small, according to the theory of Riemann integration. It means that integrable function has to be almost continuous. This theory can be applied to describe integral state of functions which have finite discontinuous points at most. But when we use this definition to describe the integrability of some functions which have infinite discontinuous points, a new problem arises. For example, Dirichlet function is not Riemann integrable on any interval. What is the integrable condition of this kind of functions? This encourages mathematicians to set up a new context, in which the integrable condition can be applied to broader cases. With a variety of further research on mathematical analysis, the theory of integration is also developed deeply. In particular, the research result of point set measure theory done by P. Jordan and F. Borel had revealed the relation between measure and integration. But the contemporary system of measure and integration most widely used was accomplished by H. Lebesgue. The theory of Lebesgue integration not only implies the outcome of the theory of Riemann integration, but also overcomes its limitations. According to Riemann, in order that a positive function $f(x)$ on $[a, b]$ is integral, the amplitude of $f(x)$ on most part of the intervals Δx_i have to be small enough. However, most of oscillation function is not integral according to the above definition. So Lebesgue introduce a new division, i.e. $\forall \delta > 0$, let

$$m = y_0 < y_1 < \cdots < y_{i-1} < y_i < \cdots < y_n = M,$$

where $y_{i-1} - y_i < \delta$, $i = 1, 2, \ldots, n$, m, M is the lower and upper boundary values of $f(x)$ on $[\mathrm{a}, \mathrm{b}]$ respectively. Let

$$E_i = \{x : y_{i-1} \leq f(x) \leq y_i\}, \quad i = 1, 2, \ldots, n.$$

Thus, the amplitude of $f(x)$ on E_i is not greater than δ. Compute

$$|I_i| = y_{i-1} \times |E_i|,$$

and

$$S = \sum_{i=1}^{n} y_{i-1}|I_i|.$$

S is a approximation of the integral of $f(x)$ on [a, b]. If $\lim_{\delta\to 0} S$ exists, and let $I = \lim_{\delta\to 0} S$, then $f(x)$ is Lebesgue integral on [a, b], noted as I.

Of course, in order to satisfy the condition of Lebesgue integral, the point sets by division $E_i (i = 1, 2, \ldots, n)$ have to be measurable. Obviously, whether $f(x)$ is integral is up to its property. In fact, we need point set

$$E = \{x : f(x) > t\}$$

to be measurable, for any $t \in R$, i.e. if $f(x)$ is a measurable function, then it is Lebesgue integrable.

Here we have a new context, in which the integrable condition of function is not only applied to functions which are "almost" continuous or bounded monotone, but also applied to measurable functions which have infinite discontinuous points. As is well known, all continuous functions are measurable and all measurable functions are Lebesgue integrable, so all of Reimann integrable functions are Lebesgue integrable. In some sense, Riemann integral is a special case of Lebesgue integral. In the above example, it is easy to see that the context of Lebesgue integral (denoted $\mathrm{C_L}$) is subsumes with the context of Riemann integral (denoted $\mathrm{C_R}$). The dynamic development process of our understandings of function integrability is the recontextualization from $\mathrm{C_R}$ to $\mathrm{C_L}$. Obviously, $\mathrm{C_L}$ is subsumes with $\mathrm{C_R}$. This means that although the recontextualization is dependent on mathematicians' pragmatic purposes in different contexts, the relation between different contexts is subsumes.

In the process of our understanding of mathematics, mathematicians make new conventions depending on current pragmatic purposes in new context when new problems arise. But this process is a continuous recontextualization. It ensures that we can find the answer to new problems by expanding the domain of mathematics in new contexts.

5. Conclusion

"From the view of contextualism, truth is no longer understood as the result of scientific exploration. It is not purely subjective or purely objective, but is to be understood as a scientific goal, a contextualized concept of the unity of subjectivity and objectivity. This kind of truth theory can reveal the diverse nature of truth from different perspectives."[3] In mathematical theory is as the same as in scientific

[3]Cheng, S. and G. Guo (2007), Truth Theory of Contextualism, Philosophical Studies, (5), 73–78.75.

theory, syntax, semantics and pragmatics are constantly evolving; both mathematical truth and scientific truth can be viewed only in dynamic context. In this sense, contextualism is a reasonable option to provide a satisfactory uniform truth theory for mathematics and science.

References

[1] Benacerraf, P. What Numbers Could Not Be [J]. *The Philosophical Review*, 1965, (74):47–73.

[2] Benacerraf, P. Mathematical Truth [J]. *The Journal of Philosophy*, 1973, 70(19): 661–679.

[3] Cheng, S. and G. Guo (2007), Truth Theory of Contextualism, *Philosophical Studies*, (5):73–78.75.

[4] Guo, G. and S. Kang, Contextual choice of Contemporary philosophy of mathematics and its significance [J]. *Philosophical Studies*, 2006, (3):74–81, 129.

[5] Guo, G. On Context [J]. *Philosophical Studies*, 1997, (4):50–51.

[6] Guo, G. The boundaries of context and its significance [J]. *Philosophical Studies*, 2009, (2):94–100.

[7] Rorty, R. Objectivity, Relativism and Truth [M]. Cambridge: Cambridge University Press, 1991:94.

[8] Zhou, Minqiang. Real Function Theory. Beijing: Peking University Press, 1998:7–8.

[9] Schlagel, R.H. Contextual Realism [M]. New York: Paragon House Publishers, 1986: xxxi.

UNDERSTANDING AS INTEGRATION OF HETEROGENEOUS REPRESENTATIONS[1]

SERGIO F. MARTÍNEZ

Instituto Investigaciones Filosofícas,
Universidad Nacional Autónoma de México,
sfmar@filosoficas.unam.mx

The search for understanding is a major aim of science. Traditionally, understanding has been undervalued in the philosophy of science because of its psychological underpinnings; nowadays, however, it is widely recognized that epistemology cannot be divorced from psychology as sharp as traditional epistemology required. This eliminates the main obstacle to give scientific understanding due attention in philosophy of science. My aim in this paper is to describe an account of scientific understanding as an emergent feature of our mastering of different (causal) explanatory frameworks that takes place through the mastering of scientific practices. Different practices lead to different kinds of representations. Such representations are often heterogeneous. The integration of such representations constitute understanding.

1. The preeminence of knowledge as the main epistemic aim has been often formulated in the philosophy of science by saying that either understanding is a mere subjective phenomenon, or else it is subordinated to the establishment of factive knowledge.[2] Hempel, for example, thought that understanding could only be a subjective (psychological) phenomenon, and thus, that it was not apt to play a role in the philosophy of science (Hempel 1965). To the extent that understanding required making reference to persons and concepts understanding was a pragmatic concept with no place in epistemology (and in the philosophy of science). One of the most influential critics and followers of Hempel, Wesley Salmon, argued that many scientific explanations did not fit Hempel's model, to the extent that the identification of causal structures articulated in mechanisms was required to understand the force of explanations. Salmon considered that such identification of mechanisms was

[1]The research for this paper has been carried with support of grant CONACYT 133345.
[2]That knowledge is factive means that it is impossible to know that p unless "p" is true. This is the sense in which knowledge is tied-down to the truth. If one says that knowledge is not factive one claims that, for example we can have knowledge of ideal gases, knowledge that an ideal gas with a given molecular weight has a given property, for example, eve though we accept that there are no ideal gases and that the proposition that an ideal gas with such molecular weight has a given property does not correspond to the facts, but it generates explanation.

"the key for our understanding of the world" (p. 160, 1984). Thus, understanding for Salmon is no more a mere subjective phenomena, but ultimately is an expression of the way Salmon thinks factive knowledge is structured. Understanding is objective to the extent that the paramount epistemic aim is the identification of causes.

This is a rather common view in the philosophy of science. Peter Lipton suggests something similar (Lipton 2004, p. 30). In this view, understanding can be taken to be something like the horizon of the causal structure of the world. That such horizon is an important dimension of understanding is no doubt correct, but it certainly does not exhaust what we try to capture in an epistemology that takes understanding as central. As Hempel was already aware, explanations of laws do not relate them to facts, but to other laws. Hempel suggested that explanations of laws consisted in situating the given law with respect to other laws, in a way that cannot be modeled as a deductive relation. Nor it reduces to processes of identification of causes.

Michael Friedman has proposed a view of understanding that picks up on Hempel's idea and overcomes some of the shortcomings of the causal view of understanding. For him understanding is related to the reduction of the number of phenomena that we accept as independent facts (Friedman 1974). Friedman shows that the idea of independence of facts is a useful and productive notion in the context of some important physical theories, but the point for us is that the unification that generates understanding for Friedman depends on some a priori structure implicit in the mathematical formalism of the theories in question. Understanding is non-factive but it is restricted in any case to the sort of theories in which such sort of formalism is relevant. More importantly for us now, it does not seem to capture some notions of understanding that are common among scientists and that are related with the value attributed to the integration of explanations in a non-foundationalist manner (see below for an elaboration of this point).

2. In contemporary epistemology there are also several attempts to characterize understanding as a central epistemological concept.[3] A detailed proposal is that of Kvanvig 2003. For Kvanvig understanding is a mental state constituted by beliefs that have to be true. The difference with knowledge according to him is that understanding is "the internal "seeing or appreciating of explanatory relationships in a body of information" (p. 198, 2003). This idea is formally close to the non-causal unificationist characterization of understanding, like Friedman's. They do not have to deny the heterogeneity of explanatory resources (causes in particular) but one way or another, they have to rely on the unificatory power of some (mathematical)

[3]I will proceed without apologies and hope that I will be vindicated once the reader understands the sense in which understanding is not tied down to the truth, but nonetheless is indirectly supported by the facts.

models in one case or some "internal seeing" in the other, to carry out the integration of resources and the desired characterization of understanding.

Whereas Friedman considers that understanding can be characterized by the sort of unification we can achieve through a reduction of independent facts, Kvanvig thinks that understanding can be spell out as some sort of coherence of a proposition to a whole body of information. If we say, I know that $2 + 2$ or, we know that the Romans conquered most of Europe; we tend to think that we know an individual proposition. But if we say, I understand $2 + 2$ or, I understand that the Romans conquered most of Europe, it suggests that what we are doing is somehow fitting the proposition into an integrated and coherent body of information.

The idea that understanding involves situating, or making coherent a given claim as part of a more comprehensive whole is a widespread idea and we agree that this is an important component of a fruitful epistemological concept of understanding. But as with independence of facts, something else have to be said in order to capture the non-aggregativity that seems to distinguish understanding from knowledge. Coherence is not enough. False theories are notoriously composed of coherent beliefs. Alchemy was a sophisticated set of coherent beliefs. Kvanvig is able to overcome obvious objections to understanding as coherence to the extent that he relies on the factive character of understanding. For example, Alchemy does not endow understanding to a proposition because the beliefs in question are largely false.

But overcoming the objection in this way excludes precisely the sort of possibility we are interested in bringing to the fore, it leaves out the the way in which coherent bodies of beliefs are the expression of constraints and standards articulated (often implicitly) in practices.

3. Understanding and the Management of Perspectives. Clearly understanding could not be identified with knowledge of the conjunction of true propositions constituting the body of beliefs with respect to which we understand a given proposition, the proposal that would be natural to make if we take understanding to be factive. Too often this body of beliefs include a false belief. That is why Kvanvig suggests that the organic feature of understanding is related to a conjunction of *central* propositions that are true, even if some peripheral propositions are false. But this suggestion leaves out of the question what are the norms that allow us to distinguish central from peripheral propositions, and more to the point, it fails to capture important uses of the concept of understanding in science. Consider the role played by diagrams in Euclidean Geometry. It seems that if we are willing to say that diagrams play a non-reducible (to propositions) epistemic role, then it is clear that diagrams have to be characterized as something more than conjunctions of propositions (see below). Analogously, state-charts (that are increasingly used in computer science, molecular biology, robotics, economics and other fields of knowledge) play an important role promoting the use of epistemic standards across disciplines, their role is not characterizable as promoting or identifying a core set of propositions, however.

For another example, take the sense in which Wimsatt argues for the importance of false models:

> Scientific models are often as useful for the false assumptions they make as for what truths they might embody. One often designs a false model deliberately for use as a template to compare with data when one is interested not in how well it fits the data, but in the form of the residuals-where it does not fit, how and why. (Wimsatt 1987, 2007 chapter 6).

Wimsatt shows how models of blending inheritance can be used to show that an important objection to group selection (as a significant evolutionary force) starts with the assumption of a strong form of blending population that is equivalent to the assumption that there is no group structure. It is then no surprise that this assumption allow us to derive the consequence that group selection is not a significant evolutionary force. The general point that is relevant for our present discussion is that a false model can help us explain better the explanatory scope of a model, and in that sense it provide us with understanding.

Notice that a false model does not need to be (reducible to) a false belief. An image, a chart, or a diagram, are able to generate understanding beyond what true beliefs can, but they are not (necessarily) false beliefs. This is related to two interesting features of understanding that seem to be epistemically relevant.

The first one is that, as many philosopher, from Hempel to Kvanvig acknowledge, understanding seems always to involve a reference point (which is always some sort of organized whole, relative to which understanding takes place). We might praise the sophistication of a newspaper in handling the history of china, whereas a specialist might find too many errors and wrong characterizations of a given episode. We can say that a child in primary school has an admirable understanding of Darwin's theory, even though we would not consider such an understanding sufficient to pass an examination at the university level. We might find a proof of a theorem quite satisfactory in secondary school, but unacceptable in a course in undergraduate mathematics, and we might find that a proof based on a diagram is accepted in a course en theoretical physics, but the proof is not be accepted as a proof in a course of mathematical analysis.

The second feature that seems relevant for discussing the difference between knowledge and understanding is very well expressed in a recent article: "the hallmark of scientific understanding is the reduction of a natural phenomenon to simpler units (Regev A. Shapiro, E. Nature/vol. 419/26 sept. 202). Understanding is often associated with some sort of reduction, the sort of reduction that promotes intelligibility (not the sort of reduction aiming for foundations).[4] It is a sort of reduction

[4]One can distinguish two senses of reduction. A fundamentalist and a non-fundamentalist type. Fundamentalist reduction assumes that theories describe layers of reality that are neatly separated and are non-controversially (objectively) identifiable. Reduction consists of explaining away less fundamental layers of reality in terms of more fundamental ones. Non fundamentalist reduction

that often, as the quote suggests, is associated with finding some basic explanatory units covering a wide variety of phenomena. As a matter of fact, the aim of Regev and Shapiro is to suggest that the lack of an appropriate abstraction from which "the mountains of knowledge about the function, activity and interaction of molecular systems in cells" could be made intelligible is not allowing the sort of understanding that accompanies successful science. They suggest that such abstraction can be found in the 'molecule-as-computation' abstraction. Insofar molecular cell biology embraces such abstraction, they hope, the possibility of sharing, comparing and criticizing claims to knowledge would be simplified and consensus reached faster.[5] Understanding involves an standpoint, but such standpoint is not factive, it requires locating the appropriate abstraction.

Notice that both features are closely related. *The search for "simpler units" is not the search for foundations but the search for the right sort of understanding-promoting perspective.* The simple units are simple in the sense that are sufficiently abstract to serve as a common ground for a wide variety of explanations that are then seen as related to each other, and thus can ground further understanding. DNA-as-string and "ideal gas" are such sort of abstractions. Their value lies not in being true, but in providing relations among the different perspectives in such a way that a robust characterization of entitlements to inferences can be orchestrated.

Lest the reader think that such account of understanding plays a role only in some relatively strongly mathematical areas of science it might be worth recalling that for example Rudwick (in Rudwick (1985)) shows the crucial importance, that for the advancement of geology, had the invention of a series of techniques for the graphical representation of different aspects of the Earth. He convincingly shows that the consolidation of geology as a body of knowledge generating understanding owns a lot to the development of a visual language (charts, diagrams and other visual resources) that eventually made possible the representation of the history of the dynamics of geological strata; maps and other techniques of graphical representation cannot be seen as mere auxiliary tools used as illustrations of theory; rather they have to be seen as constitutive elements in a new kind or argument. Further more, important studies aiming to establish the spatial and visual origins of logical and mathematical reasoning provide very general kind of evidence for such a view. A related but independent sort of evidence comes from studies suggesting that reasoning in space (visuospatial reasoning) precedes language both philogenetically and ontogenetically (see Tversky 2005, for example). Such idea suggest that the

---

does not assume a neat layered reality, and considers that reduction is more a matter of finding appropriate abstractions allowing heterogeneous representations of collective phenomena, which in turn contribute to understanding of a complex reality (that is up for grabs). As Goldenfeld and Woese put it: "...the molecular reductionism that dominated twentieth-century biology will be superseded by an interdisciplinary approach that embraces collective phenomena" (Goldenfeld and Woese 2007).

[5]Regev and Shapiro have in mind abstract computer languages such as Petrinets, State-charts and other similar proposals. They claim that such languages, developed for the specification and study of systems of interacting computations can be used to represent biomolecular systems.

role of charts and diagrams and other representational tools are not dependent on (or reducible to) language.

4. Now let us go back the question posed above, how can false beliefs generate understanding? Catherine Elgin suggests the following answer: idealizations are designed to highlight subtle matters of fact. And they do so by exemplifying features they share with the facts (see Elgin 2007 and 2009). A paint sample card can exemplify a color or a rectangle; it depends on the function the card plays in a given context. In a classroom it might be taking as exemplifying a rectangle, but in the context of looking for the color we want to paint the kitchen the card exemplifies its color. The cognitive contribution of an exemplar consists in the exemplification of selected features, and thus, anything that exemplifies exactly those features can, in a suitable context, make the same contribution (see Elgin 2009). It is not difficult to see how this account can be extended to scientific idealizations. The ideal gas (according to a theory) exemplifies features that real samples share, and to that extent can be considered to make the same contribution to the phenomena in question, even though we cannot actually separate the contribution of the real system in (or from) the real environment. This seems to be part of what is non-factive understanding about, but it cannot be all. It makes the epistemic status of idealizations parasitic on theories that make sense of the facts. Elgin sees that, and thinks that is as it should be. If those theories are overthrown we lose our reason to accept the idealization they contain.

But this does not seem quite right. Often in the history of science idealizations far outlive the theories in which they were born. "Electron" for example, is a term used by many different theories, many of which are considered false nowadays. The idealization of an ideal gas is another example. Such idealizations were born in theories that have been long discarded (at least to some extent). Elgin could argue that the idealizations are useful insofar we accept theories that support them. One could argue that statistical mechanics support the idealization of an "ideal gas". However, it is not obvious what this support amounts to. And in cases like the idealization of "electricity as a fluid" there is no specific theory in) contemporary physics supporting the idealization. The idealization is supported by the analogy or water and electric circuits. In other cases, techniques or methods and not theories are the (main) support of successful idealizations in science. Take as an example the work of Carl Woese (and collaborators) leading to a revolution in our understanding of evolution (see Goldenfeld and Woese 2007). This is a clear case in which a set of technologies (of sequencing) have an impact on basic concepts like genome and species. Finally, as pointed out above, even a false idealization (unambiguously associated with a paradigmatically false theory), such as Darwin's idealization of "blending inheritance", is a powerful tool for understanding.

Idealizations are not as parasitic on theories as Elgin seems to think. To see what idealizations can be beyond its relation to specific theories let us consider again the proposal of Regev and Shapiro. Their idea is that '*molecule-as-computation*' would point to the properties that are relevant to their function in the explanations we seek (in the explanations that would advance understanding). They think that this is the right abstraction to the extent that this abstraction can account for the key properties of *phenomena characterized by heterogeneous representations* in different areas of research in molecular cell biology. Why it is thought that the idealization of molecule-as computation is the right abstraction is because it is claimed that has the capacity to integrate heterogeneous representations.

But this would not happen by sheer "convention", or by making the idealization parasitic on a specific theory. In analogy to the way in which a cultural context makes clear that paint sample cards are intended to be used as exemplification of color, a scientific context, *which would need to be painstakingly constructed as supported by specific scientific practices (techniques, methods, models, and so on)*, would provide that conditions in which the exemplified properties would fit the abstraction (by projecting the structure of the idealized abstraction into the new field). For example, if the abstraction is to be of any use it should project the fundamental distinction between two levels of description for behavior: implementation (how the system is built) and specification (what the system does). Such distinction, as an expression of the given scientific context, would guide the ascription of biological function in a way that would constraint possible ambiguities. The 'molecule-as-computation' abstraction would thus function like the cultural context that in the case of the paint sample card makes plain to us its (paradigmatic) function.

The main claim of this paper can now be more clearly formulated. Scientific understanding can be best characterized as the integration of heterogeneous representations. The remainder of the paper is a first attempt to explain this claim. The first thing is to get clear on what are heterogeneous representations.

5. Heterogeneous representations are different representations of a thing or process that cannot be seen as aggregative descriptions, that is, they cannot be seen as representations of parts which can be seen as adding to a complete representation. If we represent a thermodynamic process using phenomenological thermodynamics we provide one sort of representation. If we characterize the same process using statistical thermodynamics we have a different representation of the same process. We might think that the phenomenological representation is subordinated to the statistical-thermo-dynamical one (or viceversa), in which case the representations are not heterogeneous. But if we think that both representations stand on its own, or at least, that both representations contribute to our understanding of the process in question (in a way that would need to be elaborated, of course), then these two representations would be heterogeneous. In this case, any one sort of representation by itself cannot capture the reality of the process in question. Of course, in order for this possibility to be taken seriously the assumption that there is, at least in principle, a single universal representational system has to go.

In theoretical contexts there is a tendency to think that different representations of the same process can ultimately be explained away and fit into a basic representation, in other words, that the heterogeneity of representations is a transitory state of our knowledge. once we take seriously the way in which knowledge is organized in practices this is a highly questionable assumption.

Consider the example of the design of a computer chip.[6] There are many different type of factors that might go into its design considerations. Engineers have identified important dimensions of design, the point of view of control, logical gates, and of timing, among others. We could think that such way of approaching the problem answer to purely pragmatic considerations, and that radically different ways of representing the different dimensions which require attention in the design could be develop. This does not sound quite right. At least for engineers, such dimensions of design (of a chip) seem to be factive. The problem is addressed by using three separate representational systems: state-charts for the representation of control, circuit diagrams for the representation of gate information, and timing diagrams for the representation of timing. Each of these systems represents some structural features to which they are homomorphic.

But the same sort of mark represents different things in the different systems. For example, certain physical relationships among lines represent relationships among various aspects of the operating chip. But lines and relationships among lines have a different interpretation in the three different representational systems. In one system of representation a line represents wiring connections, in other possible transition between states, and in the other the value on a wire through time.

We might think that the need for using different systems of representation is a consequence of our cognitive limitations, and in particular a consequence of the sort of reasoning processes we are able to perform. This might be right, but this does not need to imply that the heterogeneity of representations is a mere subjective appearance. There is at least one sense in which the different representational systems cannot be part of one single system of representation. Suppose we want to translate one representation into the other. This would require finding a common abstraction for the different representations. But each different representation involves different levels of the granularity of information. The sort of information included in timing diagrams is of a sort that is not needed for the state diagrams laying out the algorithm being computed. We can say that even if cumbersome, the translation could be carried out. But what is crucial to establish the reality of the heterogeneity of representations is that in any case there is no unique translation between diagrams (see Fisler 1993). There are any number of circuits that will implement a state machine, and any number of timing diagrams that will be compatible with those circuits. At most we can aim to establish consistency of representations. But the heterogeneity of representations stands unquestioned.

The heterogeneity of representations is often related to availability of different abstractions and different sorts of representations-cum-reasonings. Even in mathematics, often the solution of problems requires the integration of different types

[6]This example is given in Barwise 1993.

of representation. So for example, Aaron Sloman shows in an article about mental diagrams that the way in which types of abstraction and types of reasoning combines is far from simple. Depending on the abstractions used we might emphasize or de-emphasize metrical or topological reasoning, differences that might have consequences for the way problems are framed and solutions found. He finds that often one requires to combine abstract spatial representations with abstract non-spatial representations (arithmetical and logical) in order to be able to transform possible solutions at one level of abstraction (purely spatial or topological) to lower-level solutions (with metrical information) by a sort of trial and error adding of details that would lead to a clear grasp of one particular solution. Simply using concrete spatial visualization capabilities do not seem to lead us to understanding of what a solution requires. This is a fairly common type of situation. The solution to a problem is really a path through different representations that allow us to move from one level of abstraction to another. Diagrams often serves as a guide through the forest of representations which are possible or available. I want to stress that diagrams are tools that are constitutive of specific practices that norm the the way in which abstractions and reasoning interact productively.

A related way of formulating the same point can be found in Netz 1999. Netz thinks that the diagrams of Euclidean geometry can be understood as practices that through their implicit norms promote a community of what we would end up identifying as Euclidean geometers. In particular such norms constrain the boundaries of what is taken to be a good inference, and to that extent diagrams are vehicles for logic, for the sort of logic that regulates the thematic inferences distinctive of Euclidean Geometry: "the diagram is not just a graph, in the sense of graph theory. It contains at least one other type of information, namely the straightness of straight lines; that points "stand on a line" is constantly assumed on the basis of the diagram (p. 34, Netz 1999). This assertion is closely related with our claim that the right abstractions, Euclidean diagrams in this case, are not subordinated to propositions, on the contrary, Netz tells us that "propositions are individuated by their diagrams". To the extent that diagrams individuate propositions, diagrams play a role in identification of what is a good inference and thus in what is considered a proof. This is a way of suggesting that if we take seriously the importance in our cognition of visuospatial reasoning (as a sort of reasoning different from and not reducible to propositional reasoning) we are led to recognize the centrality of diagrammatic reasoning in achieving understanding.

More generally, understanding results from the integration of heterogeneous representations arising from the mastering of tools with representing capabilities (models, diagrams, charts). The tools and the representing capabilities are normed and form part of specific practices.

References

Anderson M, Meyer B, Oliver P. eds *Diagrammatic Representation and Reasoning, Springer, 2002.*

Allwein G. and Barwise, J. e. (1993). *Working Papers on Diagrams and Logic*. Bloomington, Indiana University.

Bairwise J. 1993, "Heterogeneous Reasoning", in Allwein and Bairwise 1993.

Elgin C. Understanding and the Facts Philos. Stud 2007, 132: 33.

Elgin C. Is Understanding Factive? in Haddock A, Millar A, Pritchard D, Epistemic Value, Oxford University Press.

Kvanvig, J. The Value of Knowledge and the Pursuit of Understanding, Cambridge UP.

Fisler K. 1993 A diagrammatic logic for reasoning about Asynchronous Hardware. In Allwein and Barwise 1993.

Friedman M. 1974 Explanation and Scientific Understanding, J.of Philosophy, 71(1): 5–19.

Goldenfeld and Woese 2007 *Nature* vol. 445 January 2007.

Goodman, N. (1976). *Languages of Art*, Hackett.

Grialou P. Longo G. Okada M. eds. 2005. *Images and Reasoning*, Keio U. P.

Hempel 1965 Aspects of Scientific Explanation and other Essays. New York, Free Press.

Holyoak K, Morrison R, 2005, eds, The Cambridge Handbook of Thinking and Reasoning, Cambridge U. P.

Lipton 2004 Inference to the best Explanation, London, Routledge.

Netz, R. (1999). The Shaping of Deduction in Greek Mathematics, Cambridge U. P.

Regev A. and Shapiro E. 2002, "Cells as Computation" *Nature* vol 419, 26 September 2002.

Rudwick, M. (1985). *The Great Devonian Controversy*, University of Chicago Press.

Salmon 1984. Scientific Explanation and the Causal structure of the World, Princeton.

Sloman A. Diagrams in the Mind? In Anderson M. *et al.* 2002.

Tversky B. 2005a. Visuospatial Reasoning, in Holyoak and Morrison eds. 2005.

Tversky B. 2005b, Some ways images Express and promote Thought, in Grialou *et al.* 2005.

Wimsatt, W. (2007). Re-engineering philosophy for limited beings: Piecewise approximations to reality., Harvard U. P.

Wimsatt 2002, Using False Models to Elaborate Constraints on ProcessesL Blending Inheritance in Organic and Cultural Evolution. J. Philosophy of Science 69, pp. S12–S24.

Wimsatt 1987. False Models as means to Truer Theories, in M. Nitecki and A. Hoffman, Neutral Models in Biology. London: Oxford U. Presss, pp. 23–55.

APPROXIMATION OF LAWS

ILKKA NIINILUOTO

Department of Philosophy, History, Culture and Art Studies
University of Helsinki, Finland

Approximation of laws is an important theme in the philosophy of science. If we can make sense of the idea that two scientific laws are "close" to each other, then we can also analyze such methodological notions as approximate explanation of laws, approximate reduction of theories, approximate empirical success of theories, and approximate truth of laws. Proposals for measuring the distance between quantitative scientific laws were given in Niiniluoto (1982, 1987). In this paper, these definitions are reconsidered as a response to the interesting critical remarks by Liu (1999).

1. Kepler, Newton, and Approximate Explanation

Johannes Kepler discovered his first two laws about planets in 1609. The first law states that the orbit of each planet is an ellipse with the Sun at one focus. The second law states that a line joining a planet and the Sun sweeps out equal areas in equal times. Kepler's Third Law was discovered ten years later. According to this Harmonic Law, the square of the orbital period of a planet (t) is directly proportional to the cube of the semi-major axis of the planet (a), i.e.,

$$t^2/a^3 = \text{constant for all planets.} \tag{1}$$

Here a is equal to the average distance of the planet from the Sun.

For the empirical data available to Kepler, the law (1) was remarkably accurate. When t is given in years and a in astronomical units (a = 1 for the Earth), he could make the following calculations:

	t	a	t^3/a^2
Mercury	0.241	0.3881	0.994
Venus	0.616	0.7240	1.000
Earth	1.000	1.000	1.000
Mars	1.880	1.5235	0.9995
Jupiter	11.863	5.1965	1.0029
Saturn	29.458	9.5100	1.0089

The law (1) further predicts fairly accurate values for the later discovered planets Uranus and Neptune.

Newton's mechanics in *Principia* in 1687 was celebrated for its ability to explain Galileo's and Kepler's empirical laws. William Whewell argued in 1837 that Kepler's laws were merely "formal rules", but Newton gave a "causal law" with his Universal Law of Gravitation (see Whewell, 1984, p. 59). Three years later he praised the unifying and simplifying power of Newton's theory which explained all of Kepler's three laws, "although no connexion of these laws had been visible before", and also surprisingly accounted for the fact known as the precession of the equinoxes (see Whewell, 1968, p. 153).

Empiricist philosophers accepted the explanation of laws as an important task of science. In his *A System of Logic* (first edition 1843), John Stuart Mill stated that "a law of uniformity in nature is said to be explained when another law or laws are pointed out, of which that law itself is but a case, and from which it could be deduced". However, in 1907 the French instrumentalist Pierre Duhem, who feared that the aim of explanation would subordinate science to metaphysics, criticized the Newtonian method:

> "The principle of universal gravity, very far from being derivable by generalization and induction from the observational laws of Kepler, formally contradicts these laws. If Newton's theory is correct, Kepler's laws are necessarily false." (Duhem, 1954, p. 193.)

Karl Popper welcomed this argument in his campaign against induction. In a lecture in 1948 he pointed out that "Kepler's laws cannot be deduced from Newton's", but can be obtained only *approximately* by making "the assumption that masses of the various planets are negligible compared with the mass of the sun" (see Popper, 1972, p. 357). In a paper in 1957 Popper stated that the law (1) can be obtained as a good approximation by making false assumptions. Newton's theory unifies Galileo's and Kepler's, but "it corrects them while explaining them" (*ibid.*, pp. 198–202).

Let M be the mass of the Sun, and m the mass of the planet at the distance r from the Sun. Using the "skinny triangle approximation" $\tan\theta \approx \theta$, one can show that for a planet moving with velocity v along a nearly circular orbit the centripetal force towards the Sun is $\mathrm{mv^2/r}$. By Newton's second law (force is mass x acceleration), the acceleration of the planet towards the Sun is $\mathrm{v^2/r}$. The Law of Gravitation gives the force by which the Sun attracts the planet and the force by which the planet attracts the Sun. These forces are proportional to the product Mm divided by the square $\mathrm{r^2}$. Again combining with Newton's second law, the net acceleration of the planet towards the Sun equals $\mathrm{G(m + M)/r^2}$, where G is a constant. Putting these two equations for the acceleration together, we obtain $\mathrm{v^2 = G(m + M)/r}$. For near circular orbits the velocity v is approximately $2\pi\mathrm{r/t}$, so that $\mathrm{v^2} = 4\pi^2\mathrm{r^2/t^2}$. Hence, $\mathrm{t^2/r^3} = 4\pi^2\mathrm{/G(m + M)}$. This simplified approximate derivation shows that Newton's theory entails the conclusion

$$\mathrm{t^2/a^3} = 4\pi^2\mathrm{/G(m + M)}. \tag{2}$$

This result is still idealized, as it does not take into account the perturbations by other planets. But it corrects Kepler's original empirical law (1), since the ratio t^2/a^3 depends on the mass m of the planet, and thus is different for planets with different masses. For example, the Jupiter mass is 318 times the Earth mass. Thus, if (2) is correct, then (1) is mistaken. Still, the differences of (2) for the planets in our solar system are not large, since their masses m are small in relation to the mass M of the Sun. For example, the Solar mass is 1048 times the Jupiter mass. The law (2) entails two idealizational laws:

If the masses of two planets were approximately equal,
the values of t^2/a^3 for them would be approximately equal. (3)

$$\text{If } m = 0, \text{ then } t^2/a^3 = 4\pi^2/GM = \text{constant.} \quad (4)$$

In fact, in many textbooks Kepler's original law (1), or the consequent of the counterfactual conditional (4), is derived directly from Newton's theory by assuming that the Sun is a stable massive body whose acceleration toward the planet by gravitation can be ignored. The corrected version (2) may be useful in the application of celestial mechanics to moons of planets, binary stars, and satellites.

Carl G. Hempel, whose great merit was to formulate precise models of deductive-nomological and inductive-probabilistic explanation, referred to Duhem and Popper in his remark that "while, strictly speaking, Newton's law contradicts Galileo's, it shows that the latter is almost exactly satisfied in free fall over short distances", so that "it might be said to provide an *approximate D-N explanation* of Galileo's law" (see Hempel, 1965, p. 344; cf. Niiniluoto, 1990, pp. 37–39). Similarly, we may say that Newton's theory *approximately explains* Kepler's Harmonic Law (1), since it deductively explains (2) and the distance between the laws (1) and (2) is small. Further, if the law (2) is closer to the truth or more truthlike than (1), then this approximate explanation is also a *corrective explanation.* (This is a modification of Tuomela's (1979), p. 280, condition for corrective explanation which would require that Newton's theory is more truthlike than Kepler's Third Law.)

Comparison of (2) and (4) shows that the correction of Kepler's Third Law by Newton's theory is a *concretization* in the sense Nowak (1980): the relevant factor m, which is missing in Kepler's formula, is introduced to the law (2) (cf. Niiniluoto, 1990; Kuipers, 2000).

2. Contexts of Approximation

The considerations of Section 1 can be generalized to many important methodological contexts (see Niiniluoto, 1990). What is needed is a criterion for saying that a scientific law approximates another law. If this criterion is based on a reflexive and symmetric distance or metric, two laws which are sufficiently close to each other are *approximate counterparts*. Then we can say that theory T *approximately explains* law L iff T explains an approximate counterpart of L.

Two theories T and T' are approximate counterparts if for each law in T there is an approximate counterpart in T'. Theory T *approximately reduces* to theory T' iff T reduces to an approximate counterpart of T'.

Let E be empirical data, obtained by observation, measurement, or experimentation. The distance between laws can be modified so that it applies to pairs of empirical data. Then we can say that theory T is *approximately successful* relative to empirical data E iff T entails an approximate counterpart of E.

Let L* be the *true* law governing some phenomenon. Such laws may be qualitative or quantitative, deterministic or probabilistic, theoretical or observational. Laws are stronger than mere accidental generalizations, but philosophers do not agree how to characterize their "lawlikeness". Often laws are required to sustain counterfactual conditionals and express physically or nomically necessary connections between attributes or quantities. Thus, relative to a discrete conceptual space or a continuous state space, a law of coexistence specifies what is physically possible and what is physically necessary. A law of succession, typically a dynamic or causal law, specifies physically possible trajectories of events.

The project of defining the notion of truthlikeness or verisimilitude started with Popper's attempt to give conditions for one theory (as a deductive closed set of statements in some language) to be closer to the complete truth (in the same language) than another theory. (see Popper, 1972). It continued in the 1970s with definitions of the distance of a first-order theory T from a target h*, which could be chosen as the most informative true generalization in the given language. In a simple special case, the target is a true monadic constituent which specifies which Q-predicates (kinds of individuals) are exemplified and which are not, and the Clifford distance between constituents is measured by the number of matches in their claims. A theory T typically allows several alternative constituents (complete descriptions of possible worlds). Then theory T is *approximately true*, if its minimum distance from the target h* is sufficiently small. Theory T is *truthlike*, if all of the allowed alternatives are close to h*. (For details, see Niiniluoto, 1987.)

L. J. Cohen (1980) suggested that, instead of *verisimilitude* which may concern any kinds of truths about reality, philosophers of science should be interested in *legisimilitude*, i.e., similarity of theoretical claims with the true laws of nature. Similarly, Theo Kuipers (1982) proposed that one should solve the problem of "theoretical verisimilitude". A solution to these challenges was given in Niiniluoto (1983), where monadic constituents were replaced by "nomic constituents" which express basic alternatives about physical possibilities and necessities, and the same Clifford distance can be applied to nomic constituents as well.

An alternative approach to legisimilitude is to choose a true quantitative law L* as the target, and to specify distances between such laws on the basis of their similarity (see Niiniluoto, 1982, 1988). Then law L is approximately true iff it is sufficiently close to the true law L*. A set of laws T is approximately true if the minimum distance from the members of T to L* is sufficiently small. T has a high degree of legisimilitude if the set as a whole is sufficiently close to L*, i.e., some members of T are very close to L* and T does not include too many bad alternatives (for details, see again Niiniluoto, 1987, Ch. 11).

For the concepts of *expected truthlikeness* and *probable approximate truth* as solutions to the problem of estimating degrees of verisimilitude or legisimilitude on the basis of empirical data, see Niiniluoto (1987, 2005).

The concept of truthlikeness allows us to re-evaluate Duhem's and Popper's criticism of induction. In the late 19th century, Charles Peirce introduced the notion of *abduction* as the inference from surprising facts to their explanation. According to his formulation, abduction gives a reason to suspect that a successful explanatory theory is true. But if the best available explanation T of evidence E is approximate, as in the case of Newton's theory and Kepler's laws, we may conclude for the time being that T is approximately true or truthlike (see Niiniluoto, 1999; Kuipers, 2000).

3. Distance Between Laws

In this section, I restrict my attention to quantitative statements. In the simplest cases such singular statements attribute a quantitative property to an object, so that they can be represented as points in a state space Q generated by real-valued quantities. For illustration, assume that Q is a subclass of the real plane R^2. Then the geometrical distance between two points $x = < x_1, x_2 >$ and $y = < y_1, y_2 >$ in Q can be defined by the *Manhattan* or *city-block metric*

$$d_1(x, y) = |x_1 - y_1| + |x_2 - y_2|, \tag{5}$$

the *Euclidean metric*

$$d_2(x, y) = [(x_1 - y_1)^2 + (x_2 - y_2)^2]^{1/2} \tag{6}$$

or by the *Tchebycheff metric*

$$d_\infty(x, y) = \max(|x_1 - y_1|, |x_2 - y_2|). \tag{7}$$

With each choice of d, (Q,d) is a metric space. For comparative purposes, we obtain the following standard:

$$\text{x' is better approximation of y than x iff } d(x', y) < d(x,y). \tag{8}$$

If the truth is represented by the point x* in Q, then the notions of approximate truth and truthlikeness for points x in Q can be defined by the condition $d(x,x^*) \leq \varepsilon$, where ε is the threshold of approximation. This definition can be extended to distances d(H,x*) of connected subsets H of Q from the truth x* (see Niiniluoto, 1987, pp. 303–309).

Chang Liu (1999) has made the pertinent observation that this treatment of singular approximation is "flat" in the sense that it assumes a global metric in a uniform state space Q and thus does take into account the influence of true laws on the structure of Q. For example, a claim that the velocity of light slightly exceeds the value 300.000 km/s cannot be approximately true, since it is physically impossible by the theory of relativity. As a possible reply, one could argue that the state space Q should be a neutral framework for comparing various kinds of hypotheses independently of assumed background knowledge. As false hypotheses can be truthlike, why could not physically impossible claims be approximately true?

Perhaps a more interesting line of thought is suggested by Liu's another remark: if e.g. a linear law L is assumed to hold in R^2, the singular truth x^* is a point on L, and the hypothesis y lies outside L, then the distance of y from x^* should depend also on its closeness to L. This idea would allow that physical possibilities influence the metric structure of the state space in the same way as the distribution of masses generate non-Euclidean geometries of physical space in relativity theory. In Liu's example, the circles of Euclidean metrics around the point x^* would be replaced by ellipses with L as its major axis.

Consider now simple quantitative laws which express the functional dependence of a quantity y on another quantity x. Assume that the true connection between y and x is expressed by the equation $y = f^*(x)$, where $f^* : R \to R$ is continuous function. The relevant rival lawlike hypotheses are also continuous functions $g : R \to R$. The distance between two such functions f and g can be defined by the family of *Minkowski metrics* or *L^p-metrics* for function spaces:

$$L^p(f, g) = \left[\int |f(z) - g(z)|^p dz\right]^{1/p}. \tag{9}$$

(See Niiniluoto, 1982, 1987.) The choice $p = 1$ gives a generalization of the city-block metric (5), $p = 2$ the Euclidean metric (6), and $p = \infty$ the Tchebycheff metric (7). According to L^1, two functions are close to each other if the volume between their curves is small. L^2 requires that the average distance between the curves is small, and L^∞ that the maximum distance between the curves is small.

The values $L^p(f,g)$, for $p < \infty$, are finite, if integration in (9) is restricted to a finite interval [a,b]. For comparative purposes, we may then say that law f_1 is *L^p-closer* to g than f_2 iff for some $r_0 > 0$ we have $L^p(f_1, g) < L^p(f_2, g)$ for all intervals $[-r, r]$ with $r \geq r_0$. This guarantees that f_1 is L^p-closer to g than f_2 if f_1 is uniformly closer to g than f_2 in R (see Niiniluoto, 1988).

The metric $L^2(f,g)$ can be understood as a generalization of the Least Square Difference measure in statistics, when f is a function and g is a finite set of data points. In this way, all L^p-metrics can be applied to measure the distance between Newton's result (2) and Kepler's Third Law (1). The maximum difference for Jupiter is very small, as the mass m of Jupiter is only about 0.0009546 times the mass M of the Sun.

When the notion of legisimilitude is defined in terms of the distance $L^p(f,f^*)$ of law f from the true law f^*, a law may be close to the truth even when its values f(x) do not hit the correct value $f^*(x)$ at any particular point x. Smith (1998), pp. 82–83, who criticizes the approaches of Oddie and Niiniluoto, has failed to notice that such definitions for quantitative languages are included in the Ch. 11 of Niiniluoto (1987).

Liu (1999), who speaks about "lawlikeness" in the place of our "legisimilitude", again raises interesting further questions. In his example, a fluctuating law f_1 has values which are close to the values of the true linear law f^*, but as a law it does not even remotely resemble f^*. Another law f_2 has values at greater distances from those of f^*, but its linear graph looks more like f^*. By the Minkowski metrics (9), f_1 is a better approximation of f^* than f_2, but for Liu this is the wrong answer. This argument suggests that approximation of laws may be related to two different questions: (i) what are the values of the true law?, and (ii) what is the correct mathematical form of the true law? An answer to (ii) may be correct (e.g. the law is a linear function), but its success for (i) may be very poor. Any continuous function can be approximated by a polynomial with arbitrarily high precision (see Simmons, 1963), so that a successful answer of (i) can always be found among polynomials, but as an answer to (ii) it may fail.

One way of resolving this dilemma is to acknowledge that in many methodological contexts both questions (i) and (ii) are relevant, and the overall success of an approximation should be a balanced combination of them. Smith (1998), p. 79, has something like this in mind, when he requires that in approximation one curve is transformed to another (i) by moving points by no more than some ε, and (ii) by preserving smoothness (first and higher derivatives). In Liu's example, the function f_1 satisfies (i) but fails for (ii). Another method, familiar from statistical regression analysis, is to fix the mathematical form of the searched function, either by theoretical background knowledge or considerations of simplicity, and then find among this class of functions the one which gives the closest approximation to the observed data (cf. Niiniluoto, 1987, pp. 402–403).

Note finally that Newton's correction to Kepler's Third Law serves both goals (i) and (ii). It increases approximate truth in the sense of Minkowski distance measures (9), and as a concretization of the original empirical regularity it gives an improved mathematical formula (2) for this law.

References

Butts, R. (ed.) (1968), *William Whewell's Theory of Scientific Method*, Pittsburgh, Pittsburgh.

Cohen, L. J. (1980), "What has Science to Do with Truth?", *Synthese* 45, 489–510.

Duhem, P. (1954), *The Aim and Structure of Physical Theory*, Princeton University Press, Princeton.

Hempel, C. G. (1975), *Aspects of Scientific Explanation*, The Free Press, New York.

Kuipers, T. (1982), "Approaching Descriptive and Theoretical Truth", *Erkenntnis* 18, 343–378.

Kuipers, T. (2000), *From Instrumentalism to Constructive Realism: On Some Relations between Confirmation, Empirical Progress, and Truth Approximation*, Kluwer, Dordrecht.

Liu, C. (1999), "Approximation, Idealization, and Laws of Nature", *Synthese* 118, 229–256.

Niiniluoto, I. (1982), "Truthlikeness for Quantitative Statements", in P. D. Asquith and T. Nickles (eds.), *PSA 1982*, vol. 1, Philosophy of Science Association, East Lansing, pp. 208–216.

Niiniluoto, I. (1983), "Verisimilitude vs. Legisimilitude", *Studia Logica* 42, 315–329.

Niiniluoto, I. (1987), *Truthlikeness*, D. Reidel, Dordrecht.

Niiniluoto, I. (1988), "Legisimilitude", in I. Bolnár, A. Máté, and L. Pólos (eds.), *Intensional Logic, History of Philosophy, and Methodology*, Kézirat Gyanánt, Budapest, pp. 239–246.

Niiniluoto, I. (1990), "Theories, Approximations, Idealizations", in J. Brzezinski *et al.* (eds.), *Idealization I: General Problems*, Rodopi, Amsterdam, pp. 9–57.

Niiniluoto, Ilkka (1999), "Defending Abduction", *Philosophy of Science (Proceedings)* 66, S436–S451.

Niiniluoto, I. (2005), "Inductive Logic, Verisimilitude, and Machine Learning", in P. Hajek, L. Valdés-Villanneva, and D. Westerståhl (eds.), *Logic, Methodology, and Philosophy of Science*, King's College Publications, London, pp. 295–314.

Nowak. L. (1980), *The Structure of Idealization*, D. Reidel, Dordrecht.

Popper, K. R. (1972), *Objective Knowledge*, Oxford University Press, Oxford.

Simmons, G. F. (1963), *Introduction to Topology and Modern Analysis*, McGraw Hill, New York.

Smith, P. (1998), *Explaining Chaos*, Cambridge University Press, Cambridge.

Tuomela, R. (1979), "Scientific Change and Approximation", in I. Niiniluoto and R. Tuomela (eds.), *The Logic and Epistemology of Scientific Change*, Acta Philosophica Fennica 30, North-Holland, Amsterdam, pp. 265–297.

Whewell, W. (1984), *Selected Writings on the History of Science*, ed. by Y. Elkana, The University of Chicago Press, Chicago.

SCIENCE, SEX, AND PICTURES: REFLECTIONS ON VAN FRAASSEN'S USE OF PERSPECTIVAL REPRESENTATIONS[1]

KATHLEEN OKRUHLIK

Department of Philosophy
The University of Western Ontario

Bas van Fraassen's brand of empiricism, as articulated in his well-known constructive empiricism and his more recent empiricist structuralism, has been developed using the so-called semantic approach to scientific theories. This approach has allowed van Fraassen to espouse quite a robust form of non-realism about unobservable entities while avoiding some of the problems that contributed to the demise of logical empiricism. I shall argue, however, that in other respects van Fraassen replicates many of the serious shortcomings associated with the earlier tradition — despite the fact that his choice of language sometimes obscures those similarities. Because van Fraassen's empiricist structuralism is much less well-known than his constructive empiricism, the rather lengthy first section of this paper provides an overview and initial assessment of that position. Subsequent sections deal with the ways in which van Fraassen's very sophisticated form of empiricism not only fails to address some of the most serious shortcomings of logical empiricism but actually reinforces them. The chief focus is on the failure to address issues that arise from the historical and social contexts within which science is pursued — this despite a vocabulary rich in references to perspectivity, pragmatics, context, use, and aims.

I. Scientific Representation: Pictures and Perspectives

Bas van Fraassen agrees with the slogan "All we know in science is structure." In his 2008 book, *Scientific Representation: Paradoxes of Perspective*,[†] he attempts

[1]I am grateful (as always) to Jim Brown for useful discussions and also to Lucien Lamoureux for making me think hard about some of the issues raised in *Scientific Representation*.

[†]For a longer discussion of *Scientific Representation*, see my Critical Notice, *Canadian Journal of Philosophy 39* no. 4, December 2009, pp. 671–694. The present essay provides a condensed description of the book's central argument but picks up on potential lines of criticism that my earlier treatment mentioned only in passing. Despite the critical nature of the present discussion, I must stress that the book is a *tour de force* and that I learned a great deal from struggling with it.

to develop a form of non-realist structuralism compatible not only with his own constructive empiricism but with other forms of empiricism as well. The task is challenging, since van Fraassen's commitment to non-realism requires that he develop and defend his "empiricist structuralism" without making any assumptions (implicit or explicit) about the existence of real structure in the world. If he cannot claim that our theoretical structures correspond in *some* way to underlying real structures, then how can he be a structuralist? How can mathematical structures function as scientific *representations*, if there are no structures to represent?

Van Fraassen develops his account of scientific representation within the *Bildtheorie* view of science associated with Hertz and Boltzmann — also known as the "picture theory" of science. This way of understanding scientific representation is certainly compatible with scientific realism, but van Fraassen attempts to appropriate it for his own non-realist purposes:

> If what science gives us by way of theories and models is to be conceived as pictures, as representations, then the question is opened as to just what the relevant and appropriate criteria of adequacy are for them. An extreme view would be that a good representation of that sort is one that corresponds in every respect to what it is representing. That is not what even an extreme scientific realist would say, since there is always a good deal of mathematical artifice present.... But a different criterion is found in the varieties of structuralism in the philosophy of science that we shall go on to examine — criteria to the effect that scientific models trade for their success on resemblance with respect to structure alone. And there is a still more liberal empiricist conception according to which the base line criterion is just empirical adequacy. (198)

The history of philosophy of science is littered with failed attempts to explain precisely how we might "map" our theories onto the world. Two such attempts have special significance for van Fraassen. His previous efforts to salvage Hans Reichenbach's approach to "the problem of coordination" are finally and decisively surrendered in the 2008 book. Although Reichenbach emphasized the *conventionality* of choice with respect to the coordinative definitions that connect our theories to the world, van Fraassen argues that this was insufficient to avoid the central problem, since "the coordination is still apparently to be conceived of as possible in the absence of any previous coordination," something that van Fraassen says is impossible. (121)

A failure that is even more significant for any attempt to develop a structuralist account of scientific representation is to be found in Bertrand Russell's 1927 statement of structuralism in *The Analysis of Matter*. Russell had written that

> ...whenever we infer from perceptions it is only structure that we can validly infer; and structure is what can be expressed by mathematical logic.... The only legitimate attitude about the physical world would seem to be

> one of complete agnosticism as regards all but its mathematical properties. (Russell, 1927, 254, 270)

But M.H.A. Newman in his review of the book (1928, 140) had pointed out that if two sets have the same cardinality, their 1-1 correspondence can be used to pick out a copy in one set of *any* relational structure displayed in the other. This criticism had devastating consequences not only for Russell's specific brand of structuralism but also for other attempts to develop structuralist accounts of scientific representation.[‡]

Van Fraassen recognizes the difficulty of meeting the implied challenge. No "picture theory" of scientific representation that trades on isomorphism between theoretical structures and structures in the world can succeed. So he defines his empiricist structuralism differently. It is a view *not of what nature is like but of what science is.* "Structuralism" in this instance refers solely to the thesis that all scientific representation is at heart mathematical (238–9), and the slogan that "all we know is structure" is interpreted this way: Science represents the empirical phenomena as embeddable in certain abstract structures (i.e., theoretical models), and those abstract structures are describable only up to isomorphism.

One might wonder why van Fraassen adopts the "picture theory" at all since, at least intuitively, it seems much better suited to accounts that rely on resemblance in the form of mappings or isomorphisms to establish a reference relation between the picture and the thing that it represents. The answer is that he finds in aesthetics and in art history ideas about the nature of pictorial representation that he believes will be helpful in answering questions about science in terms of *perspectival representation.* On the view that he adopts, pictures are a subset of images, which are themselves a subset of representations. Images are representations effected through resemblance, and pictures are images that bear the hallmarks of perspectivity. In science, says van Fraassen, the hallmark of perspectivity is explicit non-commitment, since being committal in some respects precludes being committal in other respects. This claim is very important and must be unpacked within the framework of van Fraassen's more general views on representation.

Evaluation of a representation as accurate or inaccurate depends on a certain kind of context. Van Fraassen says that the question, "What does it represent?" should be understood as elliptical for a more complicated question to which an appropriate answer would have the form "Z uses X to represent Y as F." *Use* presupposes intentional activity and determines our selection of similarities and dissimilarities in deciding how to represent Y. So, he says, "there is no valid inference from *what the representation is like to what the represented is like overall*" (7). This is as true in science as in art. In science, our representations of phenomena reflect such use-related pragmatic factors as aims, purposes, and instrumentation.

[‡]Newman's criticism seems to have been largely forgotten until William Demopoulos and Michael Friedman called attention to it in the 1985 article listed above.

One of the most important new distinctions introduced in *Scientific Representation* is that between "phenomena" and "appearances". In discussing van Fraassen's constructive empiricism, we could switch back and forth between expressions like "saving the appearances" and "saving the phenomena". Not anymore, however. The two terms are now sharply distinguished by means of stipulative definitions: *Phenomena* are observable entities (objects, events, processes). *Appearances* are the contents of the observation or measurement outcomes." (8) Phenomena, in other words, possess a kind of reality; they are entities in the world. Appearances, on the other hand, are *mathematical representations* of phenomena. These appearances do not tell us what the phenomena are like; they tell us only what the phenomena *look like in a given measurement or observation set-up*. This is the sense in which appearances are *perspectival representations* (i.e., pictures) of the phenomena.

It is not the phenomena themselves that are embedded in a theory's structures; phenomena are not, after all, mathematical entities. It is only pictures of them, i.e., their perspectival mathematical representations (suitably cleaned up and idealized) that are embedded in theoretical models. It is important to note that, following the introduction of the new distinction, empirical adequacy continues to be defined in terms of saving the phenomena, *not* saving the appearances.

This distinction serves the cause of developing a non-realist structuralism as follows: Instead of thinking of scientific representation as a mapping between two structures, one mathematical and one "real", we can now think of it as the embedding of one mathematical structure in another. More specifically, scientific representation consists in embedding models of the contents of our measurement outcomes (i.e., models of the appearances) in models of our theory. There is no mapping between two heterogeneous structures, only the embedding of some mathematical structures in other mathematical structures.

This brings us to a point that is critical for understanding empiricist structuralism *and* for understanding some of the concerns raised later in this paper. The case that motivates much of van Fraassen's concern with perspectival representation (as well as other aspects of his empiricism) is the measurement problem in quantum mechanics. We cannot simultaneously measure position and momentum, or spin-up and spin-down, or any conjugate pair at the quantum level. Each member of such a pair requires a different measurement set-up. We cannot remedy this defect (as we can in non-quantum cases) by using different measurement set-ups, i.e. diverse perspectives, to create an increasingly complete representation of the phenomena in question. Quantum incompleteness is built right into our theoretical accounts.

If one does not keep in mind the special relationships among the perspectival "picture theory" approach to scientific representation, the distinction between phenomena and appearances, and the problems of quantum measurement, it is very easy to be misled by van Fraassen's language about context, perspective, pragmatics, use, aims, purposes, and choices. (There seems to be a view in some quarters that van Fraassen's account of scientific representation is particularly friendly to social epistemologies of science. This, I suggest, is quite a serious misunderstanding.)

In the end, van Fraassen will use his new distinctions to claim (somewhat shockingly, at first) that requiring the theory to save the phenomena "*is not the same* as saying that the theory must be in accord with the experimental and observational results" (289). Requiring empirical adequacy does not entail that "*what happens in the actual situation must be displayed as entirely identifiable in the theoretical model*" (305). This is how van Fraassen accommodates the quantum measurement problem, but the distinction between phenomena and appearances has ramifications for the entire program of empiricist structuralism. The relationship between appearances and theoretical models is straightforward enough; it is simply a relationship between two mathematical objects. Cleaned-up, idealized models of our data sets (i.e., surface models) are embedded in our theoretical models. The relationship between theory and phenomena, however, remains untouched by this purely mathematical relationship. This is the basis of the concern that van Fraassen raises subsequently about possible "loss of reality". Connecting appearances to our theoretical models is no problem, but how do we establish and maintain a connection with *phenomena*? Mathematical embeddings alone are not sufficient to solve the problem of scientific representation.

In chapter five of *Scientific Representation*, van Fraassen spends quite a lot of time describing the kinds of historical processes through which agreement is reached about which measurement instruments and measurement processes should be used to create the appearances of those phenomena in which we have an interest. The emphasis throughout is on the perspectival nature of scientific representation and the socio-historical processes through which coordination or linkage is established between theories expressed mathematically and the phenomena to which they are answerable. Van Fraassen argues that previous attempts to establish this linkage have failed because they made the mistake of assuming that coordinations could be created independently of the historical context in which previous coordinations have taken place. He insists, on the contrary, that there is no presuppositionless starting point for establishing coordination: "measurement procedures and theory evolve together, in a thoroughly entangled way" (184).

This would seem to be the basis of a very good solution to the "problem of coordination" — indeed the only sort of solution that makes sense. We come to our scientific investigations already immersed in our natural languages and the more regimented languages employed by existing sciences. Reference is fixed through these shared languages and their associated practices. If we are willing to use the expressions that Wittgenstein employed in the *Philosophical Investigations* after giving up on his own version of the "picture theory", we can say that our forms of life and associated language games don't dictate future referential practices, but provide the shared context within which we can decide how to go on.

Van Fraassen, however, is not satisfied with these historically and socially situated accounts of science, which he says constitute the "view from within." Despite all his emphasis on pictures and perspectivity up to this point, he claims that scientific representation in its "official formulation" is *aperspectival*" (86); it is the "view from above". Goodbye picture theory. It is perhaps pedantic to point out that a "view from above" would in any ordinary usage be perspectival and that it might be better to describe the aperspectival view as the view from nowhere or the view from everywhere, depending on one's theological preferences. Van Fraassen, however, wants to deny that any God's eye point of view is being presupposed. The important thing is that the "view from above" is premised on the possibility of transcending the historical and social context within which science is created. This (perhaps unsurprisingly) is where the "loss of reality" problem rears its ugly head. Having cut the contextual ties that bound our scientific representations to the phenomena that they are supposed to represent, how do we reestablish a connection? The "problem" is a self-inflicted injury but also the biggest difficulty confronting empiricist structuralism.

Van Fraassen has what may seem to be two distinct arguments that aim to establish the model/world relationship. One is the newer "pragmatic tautology" argument, which says that the claims that (A) the theory is adequate to the phenomena, and (B) the theory is adequate to the phenomena as represented by me (i.e., as represented by my data model) are the same. The second is the older "I am here" map argument, which uses self-location to establish a connection with abstract representational structures. Van Fraassen, however, treats these two arguments as one and the same: When I say, "The graph represents the phenomena I have been measuring," it is analogous to saying that a certain point on the map is *where I am.* (257). Whether these two versions are in fact the same argument is open to question, but they both depend on how language is *used.* They are connected by considerations of *pragmatics* rather than syntax or semantics, although it is important to bear in mind that the analysis is still essentially *linguistic* in character. It is perhaps worth pointing out that in van Fraassen's writing, the adjective "pragmatic" usually refers to a branch of linguistics rather than anything resembling American Pragmatism, with its emphasis on the close connections between facts and values, theory and practice.

Van Fraassen asks us to consider a researcher describing the results of his study of changes in the deer population in the Princeton area. He describes the data model that resulted from his conscientious observations of the deer and shows how the resulting data model (suitably idealized) fits theory T quite nicely. But there is a member of the audience who insists on asking whether 'T' *fits the phenomena.* He understands that the measurements line up with the theory, and he does not doubt the researcher's skill or integrity. He just wants to know whether the theory adequately represents changes in the deer population of Princeton.

Van Fraassen argues that for the researcher to say, "My data fit the theory, but the theory doesn't fit the phenomena" is analogous to "Moore's Paradox". The researcher cannot consistently make that first-person claim any more than he can consistently assert, "P, but I don't believe P". The alleged contradiction is "pragmatic" (rather than an ordinary contradiction) because it depends on the condition of first-person assertion.

Recall, however, that van Fraassen continues to define empirical adequacy in terms of saving the phenomena, *not* saving the appearances. So it seems we *must* say that a theory saves the phenomena (is empirically adequate) exactly when the mathematical models used to represent the phenomena can be isomorphically embedded in a model of the theory — i.e., when the *representation* of those phenomena fits the theory. This does not seem to depend in any way on conditions of first-person assertion. It should perhaps be noted here that van Fraassen occasionally switches mid-discussion from the first-person singular to the first-person plural — perhaps to suggest (implicitly) an analogy between his "pragmatic contradiction" and a Putnam-type "brains in a vat" argument. This, however, is a red herring. Acknowledging that your theory, the one that guides your research and provides models within which your data models are successfully embedded, may nevertheless fail to save the phenomena does *not* involve transcending the bounds of human experience any more than it entails a pragmatic contradiction.

The only transcendence at work here is associated with the notion of an aperspectival "view from above". The idea that scientific representation can somehow transcend (or be cut free from) the context in which measurement practices and reference relations were established is extremely problematic. The negotiation of agreed upon measurement processes linking phenomena to mathematical structures cannot be replaced by "self-location" within an aperspectival mathematical structure that has been shorn of the coordinations that made it a representation in the first place.

There are major differences between van Fraassen's empiricism and the Received View of logical empiricism. The most important are these: (1) Van Fraassen's theoretical/observable distinction is not a distinction between vocabularies. It is not a linguistic distinction at all but is instead defined in terms of observability by human beings with normal sensory apparatus. (2) And as mentioned previously, his empiricism is developed within the so-called semantic rather than syntactic tradition.

Nevertheless there are some strong similarities as well. I should like to suggest that the distinction between the "view from within" and the "view from above" reinstates a divide that is analogous in some ways to the old distinction between the context of discovery (or theory generation) and the context of justification. There are, of course, major differences between the two distinctions, but the similarity lies in the shared idea that it is possible for scientific theories to completely transcend the historical context, values, desires, purposes, and background assumptions that shaped their creation.

There are also other ways in which van Fraassen's alternative version of empiricism is similar to the earlier enterprise and may suffer from some of the problems associated with its demise. These similarities and possible problems include:

- A preference for formal languages as the medium for scientific representation — in van Fraassen's case, the languages of mathematics. The role of natural languages in establishing reference with respect to scientific representation gets short shrift in the "view from within" and disappears altogether in the "view from above". It should perhaps be noted that although all logical empiricists took the "linguistic turn", they did not all do so in the same way. Neurath in particular maintained that science is inseparable from the ordinary language in which it is done. He derided the desire for a "clean" language, free from the "clots" or "congestions" that he saw as characteristic of and essential for working languages.

- The privileging of the most theoretical parts of physics (in van Fraassen's case, quantum theory) to such an extent that these are made to function as a stand-in for all other sciences. In the Received View, various types of reductionism served to justify and bolster this privileged position.
- Insufficient attention to the diachronic dimensions of science.
- An even greater lack of attention to the "external" factors and values that shape scientific choices and affect the content of science.
- Neglect of the theory-laden character of observation.

In the following section, I shall attempt to make some of these issues more vivid by drawing on examples developed by feminist scientists and philosophers of science working in the empiricist tradition.

II. Androcentric Measures

Van Fraassen's "perspectivity" acknowledges and takes into account what might reasonably be classified as one type of bias in scientific representation; it is the kind of bias that it is introduced because one has to *decide* what sorts of measurements will be made. It is a pretty good fit with the author's epistemic voluntarism. But there are many other types of bias that also affect scientific representation but enter more surreptitiously. They are not the result of explicit decisions about measurement but of the implicit introduction of shared background assumptions reflecting contextual biases. These have traditionally been dismissed as epistemically irrelevant "external values" that will be systematically eliminated from good science through rigorous methodological hygiene.

It is this mistaken assumption that has been the subject of many feminist critiques over the past 35 years. These have taken the form of detailed case studies documenting androcentric values in science with accompanying analyses of what we should make of the pervasiveness of androcentrism. In the early days, the question was whether value-laden science is evidence of (and limited to) "bad science" or whether it is in fact characteristic of "science as usual." The widespread consensus was that the latter answer was more accurate. This was not a counsel of despair or an invitation to reject science but a call to raise the epistemic bar by systematically taking into account the role that is played by values that reflect the historical, cultural, and social contexts in which science is pursued.

This has not been a purely academic exercise. Consider cases drawn from medical research. The recognition that sexist and gendered values had led to the development of data bases that reflected only male subjects, that these data bases had been used to create and confirm theories whose models reflected those same values, and finally that these theories were then used to develop treatment regimens for all patients (male and female) led to a crisis in the National Institutes of Health and similar organizations. Randomized clinical trials and other forms of testing had to be repeated using protocols that no longer reflected the pervasive gender bias of an earlier era. The change of societal values initiated by the feminist movement promoted more empirically adequate (and less dangerous) medical science. The excellent "fit" between data sets and theoretical models had been facilitated by shared social values. And the lack of empirical adequacy went largely unnoticed because precisely those same social values shaped the interpretation of the phenomena.

Consider another example: In a chapter of *The Politics of Women's Biology* called "Have Only Men Evolved?" Ruth Hubbard quotes the ethologist Wolfgang Wickler, who unwittingly provides an excellent example of "back metaphor" in sociobiology. The quote would be very funny were it an isolated example rather than a (particularly blatant) exemplification of a pervasive pattern of scientific reasoning about the roles of males and females in reproduction:

> Even among very simple organisms such as algae, which have threadlike rows of cells one behind the other, one can observe that during copulation the cell of one thread acts as male with regard to the cells of a second thread, but as female with regard to the cells of a third thread. The mark of male behavior is that the cell actively crawls or swims over to the other; the female cell remains passive. (98)

So the circle is complete: Start with social stereotypes about active males and passive females; embed these stereotypes into evolutionary theories and into the observation protocols used to collect data so that these stereotypes are projected onto all organisms, even algae. (Note that one and the same thread of algae is male when it is active, female when it is passive.) Then use the resulting data to confirm as scientific the claim that there is a pervasive *biological* basis for the association of males with activity and females with passivity. The theory, its models, the "measurement" techniques adopted, and the resulting data sets are all informed by the same value-laden background assumptions. And it is these shared assumptions that become the basis of claims about empirical adequacy. Without systematically taking into account the social values and background assumptions that facilitate this "fit", we will fail to recognize that the theory is empirically inadequate, that it fails to save the phenomena.

This is not just a case of "perspectival" representation that is necessarily noncommittal regarding some aspects of the phenomenon. Nor is it simply an imperfect representation — one that still needs work. It is empirically inadequate, even if the mutually reinforcing theories, measurement techniques, and data sets all line up very nicely. The illusion of empirical adequacy is facilitated and supported by the pervasiveness of the shared bias.

Let us return now to van Fraassen's Princeton researcher, but let us suppose that instead of discussing the local deer population, he is instead reporting on his recent work in Africa on social structures and behaviors among certain higher primates. Standing in front of the room, he has carefully described how his conscientious observations of a specified population resulted in a data model that fits very nicely with the mathematical models of the reigning theory about primate behavior, reproductive roles, and social structure.

One of the women in the back of the room raises her hand. "I can see," she says, "that there is a good fit between T and your data model. But does T fit the actual primate population that you have studied?" The other women in the back row lean forward to hear his answer.

"Well," says the young researcher, "this is my data set. It is the result of painstaking observations, and the fit with T is excellent. I stand by both those claims and do not wish to withdraw them".

"But you know what? My mother recently sent me some really interesting work by Donna Haraway, Sarah Blaffer Hrdy, and other feminist primatologists. After reading those articles, I'm beginning to wonder whether theory T really is adequate to the phenomena in question. In retrospect, I think that using the structures afforded by the models of T may have made certain aspects of the phenomena virtually invisible while leading me and other traditionally trained investigators to assign undue significance to other aspects of phenomena — those whose representations could find a comfortable home in the models of T. I also worry in retrospect that our observations were not just 'perspectival' but also heavily theory-laden."

"I understand," continues the researcher at the front of the room, "that all scientific representation is perspectival. Really, I do; that guy van Fraassen makes some really good points. So I'm not suggesting that that adopting a specific perspective is unscientific; it is both unavoidable and essential. It doesn't follow, however, that every perspective provides a good representation of the phenomena or that choice of perspective is always an innocent act. More importantly, however, it seems that we typically observe the phenomena not simply from different perspectives, i.e., different measurement set-ups, but through lenses of diverse colors and shapes. Call it value-ladenness, call it what you will; but it must be taken into account."

"Why did I fail to observe instances of female aggression in the langur monkeys I was observing? Or *any* examples of female choice and agency? What I believe now is that my theory and its models simply made no room for such observations. I have subsequently learned (from Mom) that women in the field seem to observe quite a lot of female agency — including infanticide, promiscuity, and aggression against males. Their own experiences and background assumptions as well the alternative theoretical models they sometimes adopt make room for such behaviors, whereas my models did not."

"It's partly that men and women bring different experiences to their field work. But it's also because most of us have been educated in the Darwinian tradition in which it is male competition that drives the evolutionary engine. So, our theories about primate social organization focus on male behaviors. These theories and their models provide our background assumptions. They shape our expectations, the perspectives we adopt, and thus the observations we make, the experiments we perform, and the data that we record as relevant."

"If I did not observe instances of female agency but have subsequently learned that female primatologists have documented such behaviors among the same population I was studying, why is that? My vision is 20/20. I have a PhD from a top-tier program, and I keep meticulous records. Why did I record not one single instance of female agency relevant to the monkeys' social structure or reproductive success?"

"It's true that the perspective from which I made my observations dictated what I would record. But the theory that provided the models that informed my research was itself bolstered by other data models that became canonical in part because they 'fit' with the wide range of value-laden background assumptions that mediate the distance between data and hypothesis.[§]"

"It's not simply that most research in primatology has been based on background assumptions about the primacy of male struggle in the evolutionary process but also because we have canonical *data sets* that seem to conclusively support this theoretical orientation and to fit its mathematical models. Bateman's fruit fly data seemed to show conclusively that females breed at close to capacity, with the result that differential reproductive success (and so the course of evolution) must be the result of male competition. Those Bateman data are at the very heart of sociobiological models of human behavior and social organization, and those models in turn are often cited to justify traditional gender roles."

"Just yesterday, though, Mom sent me a link to an article written by scientists who tried unsuccessfully to replicate Bateman's studies.[¶] They discovered pretty glaring deficiencies in his data and showed that his methods were incapable of establishing the intended conclusion, that 'sexual selection acted primarily on males through female choice and through male competition and profligacy in mating.' The authors ask why nobody had ever questioned Bateman's data sets before. Their suggestion is that the inaccuracies weren't previously noted because the data fit so nicely with "the reigning world-view."

"I don't know, but I am increasingly convinced that we cannot neglect contextual factors (beyond those associated with perspectivity) when assessing the observations we use to assess empirical adequacy. Perspective is important, but the larger social and historical context is perhaps even more important, since it influences not only our choice of perspective, but also model choice and even the content of our observations."

"So, dear interlocutors in the back row, the only way I can honestly answer

[§]See Longino and Doell for the classic discussion.

[¶]Gowaty, Kim, and Anderson. For a short and interesting discussion of this article, see http://www.newappsblog.com/2012/07/a-key-study-on-sexual-selection-fails-the-test-of-replication.html your question is to say: This is indeed my data set; it is the product of careful observations carried out under the auspices of T, and it fits T quite beautifully. I also recognize, however, that it is very likely that T systematically misrepresents the phenomena. Feminist critiques of the biological sciences have led me to believe that this is frequently the case. I fail to see any pragmatic contradiction in standing by both assertions."

"What a nice young man," say the women in the back row. "His mother must be proud."

III. Anthropocentric Pleasures

One the most striking features of van Fraassen's account of picture theory and its perspectival representations is the absence of any real discussion of the theory-ladenness of observation, although this topic is typically front and centre in discussions of partiality in the representation of phenomena. The neglect of the theory-laden character of observation by classical logical empiricists was one of the main targets of a host of critics and one of the chief motivations for developing accounts of science that do not neglect its historical and social settings. In introducing to philosophy of science the idea that all observation is theory-laden, Norwood Russell Hanson offered this argument:

> Objects, events, pictures are not intrinsically significant or relevant. If seeing were just an optical-chemical process, then nothing we ever saw would be relevant to what we know, and nothing known could have relevance to what we see. Visual life would be unintelligible; intellectual life would lack a visual aspect. Man would be a blind computer harnessed to a brainless photoplate.[‖]

Van Fraassen's observer must be much more than a blind computer harnessed to a brainless photoplate. His observer must be the same epistemic agent who makes decisions about model choice, about what measurements to undertake, which perspectives to adopt, what to believe. So it seems that the theory-laden and contextual nature of observation must somehow be accommodated. We have already seen that van Fraassen's move to perspectival representation was largely motivated by the quantum measurement problem, an issue that appears to be independent of concerns about theory-laden observations. I shall suggest, however, that it is not simply the case that van Fraassen does not emphasize the theory-laden character of observation but that arguments he makes in defining his own version of empiricism tend to mitigate against it.

The biological sensory apparatus of homo sapiens plays the central role in defining van Fraassen's anti-realism. Recall from The Scientific Image that the single

[‖]Patterns of Discovery, p.26. The arguments in chapter one, "Observation", provide a systematic and striking contrast with van Fraassen's views on the same topic.

aim of science is empirical adequacy and that "a theory is empirically adequate if what it says about the observable things and events is true-exactly if it 'saves the phenomena,' "- where observability is defined in terms of the unaided human senses (1980, 12). It is important for present purposes to stress that van Fraassen clearly distinguishes between "observing" and "observing that" (1980, 15). Only the latter interpretive moment ("observing that") is affected by background knowledge, expectations, etc. And only the former (i.e., observing) is involved in the definition of empirical adequacy as the single aim of science. In a later works (2000) (2002), van Fraassen discusses "literate experience", but again this concerns interpretation rather than observation. What is observable does not depend on theoretical commitments, expectations, or values but only on the unaided human senses.

In singling out empirical adequacy as the one defining aim of science, van Fraassen explicitly rejects explanation as an additional aim. "The discussion of explanation went wrong," he says, "at the very beginning when explanation was conceived of as a relationship like description: a relation between theory and fact. Really, it is a three-term relation between theory, fact, and context." As a result, he says, scientific explanation is not pure science but "a use of science to satisfy certain of our desires" in a quite specific context. Explanation, he says, is an "anthropocentric pleasure." (1980,156)**

I do not wish to argue with this but only to point out that observation and description are not very different from explanation in this respect. There can, of course, be no doubt regarding their anthropocentric nature, since it is the physiology of homo sapiens that determines what is observable. In addition, however, we make our observations in quite specific contexts, which are informed by our goals, desires, and expectations as well as our background assumptions. We try to locate our data sets in theoretical models that were chosen under similar influences. These affect our decisions to make certain sorts of observations but not others. They also affect what happens next: Which observations are recorded, and how are they recorded? Which are discarded as background noise? Among our observations, which seem salient, and which do not? All these questions are relevant to the primatology issues raised in the previous section, but it may be useful to look briefly at salience as an example.

Many of the classic feminist case studies focus on instances in which phenomena are not noticed or simply ignored by scientists, who regard them as irrelevant because they do not fit reigning theoretical models and conventional expectations. In the cases mentioned above, these were typically examples of female agency that might have an impact on the course of evolution. There is no place in our "best" models for data about such phenomena; so either the phenomena are not noticed or they are intentionally ignored. The problem lies not with limitations of anyone's sensory apparatus but with limitations of the epistemic agent - a point that the nice young man in section two seemed to acknowledge.

** *The Scientific Image*, p. 56.

It must be emphasized that drawing attention to the fact that observation and description are anthropocentric pleasures does not detract in any way from the status of empirical adequacy as the central constitutive value of science. Instead it reminds us that we must surrender all hope of achieving a view from above or a view from nowhere. And it highlights the fact that science itself is one of life's most remarkable anthropocentric pleasures.

References

Gowaty, P., Y.-K. Kim and W. Anderson (2012) "No evidence of sexual selection in a repetition of Bateman's classic study of *Drosophila melanogaster*," *Proceedings of the National Academy of Sciences* 109 (29), 11476–11477.

Haraway, Donna (1990) *Primate Visions: Gender, Race, and Nature in the World of Modern Science*, New York: Routledge.

Hrdy, Sarah Blaffer (1986) "Empathy, Polyandry, and the Myth of the coy Female," in *Feminist Approaches to Science*, ed. Ruth Bleier, New York: Pergamon, 1986, pp. 119–146.

Hubbard, Ruth (1990) *The Politics of Women's Biology*. New Brunswick: Rutgers University Press.

Longino, Helen and Ruth Doell (1983) "Body, Bias, and Behavior: A Comparative Analysis of Reasoning in Two Areas of Biological Science," *Signs*, Vol. 9, No. 2 (Winter, 1983), pp. 206–227.

Okruhlik, Kathleen (2009), Critical Notice of van Fraassen (2008), *Canadian Journal of Philosophy 39* no. 4, December 2009, pp. 671–694.

van Fraassen, Bas C. (1980) *The Scientific Image*. Oxford: Oxford University Press.

______ (2000) "Literate Experience: The [De; Re] Interpretation of Nature" Available at http://webware.princeton.edu/vanfraas/mess/literate.htm.

______ (2002)The Empirical Stance. New Haven: Yale University Press.

______ (2008) *Scientific Representation*. Oxford: Oxford University Press.

BAYESIANISM VERSUS CONFIRMATION

MICHAEL STREVENS

Department of Philosophy
New York University
5 Washington Place
New York, NY 10003, USA
Email: strevens@nyu.edu

The usual Bayesian approach to understanding the confirmation of scientific theories is inadequate. The problem lies not with Bayesian epistemology, but with a simplistic equation of the subjective, individualistic evidential relevance relation that Bayesianism attempts to capture and the more objective relevance relation of confirmation.

1. Introduction

Grant that Bayesian epistemology provides a respectable theory of rational belief revision; my question in this paper is how the Bayesian theory of belief revision is related to the confirmation relation in science.

The core of a Bayesian account of theory confirmation is typically thought to be straightforward:

> *Simple Proposal:* A piece of evidence confirms a theory for a reasoner just in case the evidence increases the reasoner's subjective probability for the theory (in accordance with the Bayesian conditionalization rule), and it disconfirms a theory just in case the evidence decreases the subjective probability.

As implied by the wording, confirmation is always relative to a reasoner: what increases the subjective probability of a hypothesis for one scientist may decrease it for another.

This bare account of the relation between Bayesian probability kinematics and scientific confirmation can be augmented in various ways. You might introduce a notion of degree of confirmation,[1,2] or a more static notion of evidential relevance that allows a piece of evidence to remain relevant to a hypothesis even after it has had its one-time effect on the hypothesis's probability.[3] These elaborations will not, however, concern me here. I want to argue that the Simple Proposal presented above is incorrect, that it must be replaced with something that better captures the distinctive features of science's confirmation relation.

2. Against the Simple Proposal

A scientist performs an experiment to test hypothesis h. The scientist's background beliefs, together with h itself, logically entail that the result of the experiment will be e. The scientist, however, incorrectly believes that the theory and background entail $\neg e$. As it happens, the experiment produces result $\neg e$, contrary to the hypothesis's prediction. The scientist, however, thinks that the experiment has produced just the result that the hypothesis foresees. Is the hypothesis confirmed or disconfirmed?

The Bayesian has two ways of thinking about this problem. Bayesian epistemology, in its classical form, incorporates an assumption of "logical omniscience": a Bayesian is supposed to be logically infallible (and is supposed to be aware of all theoretical possibilities). The classical Bayesian model cannot, consequently, be used to represent reasoning in which logical errors are made; it simply does not pronounce on the case at hand.

It is, however, widely recognized among those who favor a Bayesian representation of scientific reasoning that such reasoning does not conform to the assumption of logical omniscience. A more tolerant or "humanized" variant of Bayesianism is needed, on which, for example, logical tautologies can be attributed degrees of belief less than one.[3,4] The technical obstacles to constructing such an account are serious, but put all such worries on hold and suppose that we are dealing with this sort of logically fallibilist Bayesian theory.

A humanized Bayesianism will treat the experimenter as having a high subjective probability for $\neg e$ conditional on h (though the axioms of probability require a zero probability). Consequently, conditionalizing on $\neg e$ will boost the experimenter's subjective probability for h.[a] The humanized Bayesian judges, then, that the experimental result in the above scenario confirms the hypothesis—if the Simple Proposal is correct. (More exactly, it confirms the hypothesis for the scientist in question; it would disconfirm the hypothesis for a bystander with the same background beliefs who correctly saw that the hypothesis predicts e.)

Is this right? It is not. The way that scientists think and talk about the relation between theory and evidence—about the confirmation relation—mandates the following description of the case: the experimental result disconfirms the hypothesis, but the experimenter because of their logical error falsely believes that the result confirms it. What matters for confirmation, then, is the objective logical relation between the hypothesis and the evidence, not what a scientist takes that relation to be.

Before you begin to raise objections, let me describe two other cases that I take to raise similar problems for Bayesian confirmation theory.

Consider a scientist investigating the hypothesis *Gold is diamagnetic*. They test a specimen of what they believe to be gold, and find that it is paramagnetic, and

---

[a]I am assuming of course that the experimenter's unconditional probability for $\neg e$ is lower than this conditional probability.

thus not diamagnetic. The sample, however, is not gold but iron pyrite (fool's gold). Because the scientist is confident, and by assumption remains confident, that the sample is gold, their subjective probability for its being paramagnetic conditional on the hypothesis is very low, and certainly, much lower than their unconditional probability for paramagnetism. So their subjective probability for the hypothesis will decrease on their learning of the paramagnetic quality of the sample. According to the Simple Proposal, then, the fact of the sample's paramagnetism disconfirms the hypothesis.

This is, again, incorrect. Because the hypothesis makes no claims about the magnetic properties of iron pyrite, the observation of these properties, whatever they may be, is irrelevant to the hypothesis. The pyrite's paramagnetism neither confirms nor disconfirms the hypothesis, which has things to say only about gold. The investigator falsely believes that the hypothesis is disconfirmed, but they are wrong in this belief.

Or consider the computer program PYTHIA that is used to derive statistical predictions from the complex mathematics of theoretical particle physics. Suppose that an experimenter testing a novel theory—call it Ubersymmetry—uses PYTHIA to infer the physical probability that the theory ascribes to some particular event.[b] They have every reason to trust PYTHIA, as it has been exhaustively tested and extensively used. But the program may not be free of errors. For that matter, a computer on which the program is run may malfunction. In such circumstances, the experimenter might well come to have a false belief about the physical probability that Ubersymmetry assigns to an event. Suppose that this unfortunate train of events comes to pass. As a consequence, the experimenter falsely—though blamelessly—believes that Ubersymmetry assigns a physical probability to an event that is quite at odds with the event's observed frequency, when in fact the frequency and the probability are close.[c]

The experimenter's subjective probability for Ubersymmetry will drop; had they had the correct belief about the physical probability, it would rather have risen. Does the evidence confirm or disconfirm Ubersymmetry? It confirms it, I suggest: in scientific discourse, confirmation is a matter of what a theory actually predicts, not of what some experimenter thinks it predicts. The Simple Proposal fails to deliver this judgment.

---

[b]Because PYTHIA uses Monte Carlo methods, the inference is inductive rather than deductive even if the issue of the fallibility of the program and machine are put aside. This aspect of PYTHIA is irrelevant to the example (though of course it might be used to generate a similar scenario).

[c]In this case, then, the experimenter violates Lewis's Principal Principle,[5] which like other principles of probability coordination requires (in the kind of circumstances I have in mind here) that the subjective probability for the evidence conditional on the hypothesis equal the physical probability ascribed to the evidence by the hypothesis. The violation is blameless, indeed rational, because the experimenter has good reason to believe that PYTHIA is reliable.

3. Subjective versus Objective

What is wrong with the notion of confirmation offered by the conjunction of Bayesian epistemology and the Simple Proposal is that it is in a certain sense too subjective: it implies that a hypothesis is confirmed (for a scientist) whenever the scientist *thinks* that it is confirmed. As the examples above show, however, confirmation as it figures in science is a more objective relation than that. Whether the evidence confirms a hypothesis depends on what the hypothesis says about the evidence, not what some scientist believes—perhaps mistakenly—that the hypothesis says about the evidence.

Where does the fault lie? Not, I suggest, in Bayesian epistemology. In the cases above, Bayesian updating is doing the job it is supposed to do, capturing a sense of evidential relevance that is germane to individualistic epistemology, that is, to deciding the question of what an individual ought, given their epistemic situation, to believe. This is most obvious in the PYTHIA case, where I have given you enough of the epistemic background to make it clear that the scientist in question ought to take PYTHIA's (incorrect) judgments seriously and lower their confidence in the Ubersymmetry theory, even though to do so is to move away from the truth. But the same can be said of the other cases, too: given the scientist's belief that their sample is gold, for example, its paramagnetism provides good reason to relinquish the hypothesis that gold is diamagnetic.

The problem, then, must be attributed to the Simple Proposal, which attempts to capture the relation of scientific confirmation using an account of the individualistic evidential relevance relation. Confirmation is an evidential relevance relation, to be sure, but it is not individualistic in this sense. Science is an intersubjective, indeed a social, pursuit. For this and other reasons, it requires a more objective conception of evidential relevance, a conception on which it is possible for an individual scientist to fail to grasp the evidential significance of an observation.[d]

Two brief comments. First, that Bayesian confirmation theory is "too subjective" is a well-known concern. Such a complaint is insufficiently fine-grained, in my view; it ought to be divided in two. On the one hand, you can criticize the Simple Proposal as too subjective, as I have here, because it attempts to align the more objective confirmation relation with the more subjective relation of evidential relevance proper to an individualistic epistemology. On the other hand, you can criticize Bayesian epistemology as too subjective even as an individualistic epistemology, on the grounds that its relations of evidential relevance are entirely determined by a personal assignment of prior probabilities on which there are almost no constraints. Both objections perhaps have bite, but it is a mistake to assume that what is too subjective for science is thereby also too subjective for the individual.

Second, you might wonder whether scientists use the term "confirmation" to talk about many different evidential support relations, some more subjective and some

[d]For some other ways in which the confirmation relation is intersubjective or objective rather than individualistic, see Refs. 6 and 7.

more objective, in different contexts. Perhaps there is some flexibility in the term, but I think (although in this short paper I have not offered sociological evidence) that there is a special role in scientific discourse for a relation based on the actual rather than the fancied implications of a theory. Even if you do not think that this relation dominates all others in scientific discourse, I hope I will have persuaded you at least that it has an important place, and so that there are significant uses of the term "confirms" that are not captured by the Bayesian conception of support.

4. Beyond the Simple Proposal

To move forward, what is needed? Two things: an account of the confirmation relation that is relatively independent of individualistic epistemology, and a replacement for the Simple Proposal that gives a more sophisticated account of the relationship between individualistic epistemology and the confirmation relation. I will say no more about the former project. I do have a few concluding, schematic comments about the latter project, that is, about the proper interface between individualistic epistemology and science's confirmation relation.

The individualistic relevance relation cannot be equated with the confirmation relation. But confirmation can and will enter into an individual's reasoning, and thus into individualistic epistemology, by way of beliefs about the confirmation relation. In the PYTHIA case, for example, the drop in the experimenter's subjective probability for the theory reflects their (false) belief that the evidence disconfirms the theory.[e] These beliefs about confirmation will of course closely track beliefs about what the hypothesis says about the evidence, which may in turn track beliefs about the logical implications of the hypothesis (for example, what physical probability the hypothesis assigns to an event), beliefs about the properties of the evidence (for example, whether a specimen is gold), or perhaps even beliefs about the content of a ceteris paribus hedge.[8] The proper tracking relation might be captured by the following sort of constraint: ceteris paribus, the recognition that a hypothesis is confirmed should provide individualistic good reason to increase your subjective probability for the hypothesis.

A comprehensive individualistic epistemology, such as humanized Bayesianism, will incorporate all the forms of uncertainty described in the previous paragraph. The good news, then, is that such an epistemology should be able to interface with any reasonable theory of the nature of confirmation. The corollary of this flexibility—the bad news, for those who like things simple—is that an adequate theory of the confirmation relation cannot be read off from the correct theory of individualistic epistemology, whether it is Bayesianism or something else.

References

1. P. Milne, *Philosophy of Science* **63**, 21 (1996).

[e]The Bayesian apparatus, incidentally, provides no resources for saying that one of these facts is grounded in the other.

2. B. Fitelson, *Philosophy of Science* **66**, S362 (1999).
3. D. Garber, Old evidence and logical omniscience in Bayesian confirmation theory in Earman[9]
4. R. C. Jeffrey, Bayesianism with a human face in Earman[9]
5. D. Lewis, A subjectivist's guide to objective chance, in *Studies in Inductive Logic and Probability*, ed. R. C. Jeffrey (University of California Press, Berkeley, CA, 1980)
6. M. Strevens, *Philosophical Studies* **143**, 91 (2009).
7. M. Strevens, *Oxford Studies in Epistemology* **3**, 294 (2010).
8. M. Strevens, *Journal of Philosophy* **109**, 652 (2012).
9. J. Earman (ed.), *Testing Scientific Theories*, Minnesota Studies in the Philosophy of Science, Vol. 10 (University of Minnesota Press, Minneapolis, 1983).

THE EQUIVALENT TRANSFORMATION BETWEEN NON-TRUTH-FUNCTION AND TRUTH-FUNCTION

XIAO-LONG WAN* and MING-YI CHEN†

**Prof. & Chair in Unit of Logic and Philosophy of Science and Tech., Department of Philosophy, Huazhong University of Science and Technology, P. O. Box 430074, Wuhan, Hubei, P. R. of China.*
hwanxl@163.com.
†PhD Candidate in Logic and Philosophy of Science, Department of Philosophy, Huazhong University of Science and Technology, P.O. Box 430074, Wuhan, Hubei, P. R. of China.
cmy0223@163.com.

Unary Operator Theory (UOT) systematically researches unary operators by analyzing the unification between the formal universality of mathematical logic and the adaptability of the natural language. At first, it focuses on Special Theory of Relativity of Function (STRF) which is only useful for equivalent transformation between the non-truth-valued-function formed by non-functional unary operators and truth-valued-function formed by functional binary connectives in two-truth-valued classical propositional logic. Moreover, according to the STRF, there are new interpretations about problems both in philosophical logic and in philosophy of science which could not be solved only by the logic of truth-function in the past.

Keywords: Equivalence Non-Truth-Function Truth-Function STRF.

1. Introduction

The modern mathematical logic starting from Frege and Russell, which develops a new technology in the field of logic, has become a common theoretical foundation for information science and modern analytic philosophy. However, the classical part of modern logic seems not very well to deal with some complex problems, such as "the failure of equivalent substitution principle" which often appears in natural language, traditional philosophy, quantum theory and computer language. For example:

"I am very hungry but I have serious liver disease. I eat food and drink alcohol. I ought to eat food, and I ought not to drink alcohol".

The Logical semantics in the last several sentences is: "I eat food" is true; "I drink alcohol" is true. "I ought to eat food" is true; "I ought to drink alcohol" is false.

Therefore, there is a formal problem (T): the two compound sentences are constructed by two component sentences with the same connective "ought to". The two component sentences have the same truth-value, but their compound sentences have different truth-value.

Such problem (T) labeled as "the failure of equivalent substitution principle" had puzzled the founder of modern logic, Frege, and his followers, for a long time. Frege, and most logicians later thought that we had to use intensional logic to resolve it. It is well-known that modal logic is a typical intensional logic.

The modern modal logic, as one of the most typical and mature kinds of non-classical logics, not only has a large and complex systems in foundation, but also is widely applied in computer science, modern analytic philosophy, the interpretation of quantum mechanics and even many social sciences and humanities. It seems that modal logic can solve the problem (T). Moreover, Quine's questioning the validity of early modern modal logic stimulated the development of modal logical theories, such as the addition of the necessitation rule in syntax and the invention of possible world semantics. However, the possible world semantics does not make a modal formula correspond to a first-order formula one by one, and vice versa. Whats more, morally generalized modal logic processing will produce logical paradoxes which are difficult to solve. The friends in information science, economics, and mathematics tell me, they generally use statistical methods to solve the problem (T), because they worry about the non-total unity between modal logic and classical logic in information science and artificial intelligence.

Furthermore, both the intensional logic and extensional logic are formal logics, so any intensional logic is a formal reasoning of intensions. Thus, the further clear technology distinguishing extension logic and intensional logic is that the formal is about truth-function, while the later such as modal propositional logic is about non-truth-function. Equivalent substitution principle holds with truth-function but no longer holds with non truth-function. It is seemly why modern classical logic (or called standard logic) up to now can not effectively deal with the problem in the previous section of this article, for "ought to" is obviously not a truth-valued functional connective. That is to say, classical logic is just a truth-function logic.

A few papers are dedicated to systematically investigate into the nature of the non-truth-functions or of the non-truth-functional connectives (Quine, W. V. O. 1982, p. 8; Schnieder, B. 2008, p. 64; Hill, D. J. and Mcleod, S. K. 2010, p. 629). Even though they had studied in detail the differences between non-truth-function and truth-function or differences between non-truth-functional connective and truth-functional connective (Marcos, J. 2008, p. 215; Rescher, N. 1961, p. 1–10, Stephen, P. 2005, p. 93–105) no one recognizes the equivalent transformation between them, nor do more refer to give such transformation a tightly clear definition. Moreover, can the classical (propositional) logic truly not handle the problems with "non-truth-function operates"?

There are still some philosophers faithfully defend only effective position of classical logic and refused to acknowledge the non-classical logic correction to the classical logic. In this respect, University of Oxford philosopher Tim of West Williamson (Timothy Williamson) is the most representative of the defense, He said: "the classical logic is normally used in logic, mathematics and science, in these areas it has proven to be extremely successful.'if, bad workman blames his tools', then the classical logic and semantics is the philosophical craftsmen poor tool"[1] There are many bottom-up ways to explore the relationship between classical logic and non-classical logic, but the characteristic of this article is along a top-down approach. We intend to pursue a unified method or comprehensive theory to understand the relationship between the non-classical logics and classical logics, on the basis of formal university of classical propositional logic and its proper compliance with natural language. Since classical propositional logic is the most mature and the simple modern logic, special modal operators are usually viewed as non-truth-functional and unary operators, so the axiomatic systems of the most non-classical logics are a set of axioms of truth-functions in classical logic plus the additional axioms of non-truth-functions formed by modal operators or other non-truth-functional connectives (Kripke, A. S. 1965, p. 208; Ming Xu 2008, p. 492), we concentrate on the relations between truth-valued function connective in classical propositional logic and non-truth-valued function operate in modal propositional logic with two truth-valued in this article. The Special Theory of Relativity of Function (STRF) is invented to deal with this relationship, and it is considered as the first step of more general one, "Unary Operator Theory". Perhaps, classical logics and non-classical logics are to some extent unified in very basic respect. According to STRF, at least, we can further paraphrase modal propositional logic (and even some many-valued logics) as an "fragment", or even another sight of classical logic.

2. STRF

The summary of aforementioned section is: classical logic vs. $non - classicallogic \to extension$ logic vs. intension $logic \to truth - function$ vs. $non - truth - function \to classical$ propositional logic vs. modal propositional logic. STRF studies the equivalent transformation between the two-valued non-truth-functions formed by non-functional unary operators (connectives) and truth-functions formed by functional binary connectives in classic propositional logic only.

2.1. *Preliminary consideration*

Generally speaking, a function is a corresponding relation of two variables (or of two sets or of two formulas). For a variable x and dependent variable y both in their valued-domain, if x has a value corresponding y having a and only a value, then this y is a function of x. Truth-function is the function in which the domain of the variable and the dependent variable are all the truth valued-domain only. Non-truth-function is not that the function in which the domain of the variable and of the dependent variable is all the truth valued-domain only. It includes the

[1]Chen Bo. 2008. Thinking Deeply, Contributing Originally: An Interview with Timothy Williamson. Studies in Logic 1 (3): 79–109.

non-truth-valued function and the truth-value of non-function. That is to say, we study the non-truth-function is actually two-truth-valued non-function. However, to have the philosophy of logic corresponding to the text, we still call it non-truth-valued function.

In Table 1, $D_2 = p \vee p_1$ is composed of p and p_1, these two variables together form a truth-valued function, but D_2 is not a truth-valued function by variable p only (when p takes a definite truth-value 0, $p \vee p_1$ is not only a corresponding definite truth-value). Therefore, D_2 is a non-truth-valued function of p. This shows that at least some of the non-truth-function is actually equivalent to a truth-function in another sight. Among 16 truth-functions in Table 1, though they are all the truth-functions formed by p and p_1 together, only D_1, D_6, D_{11}, and D_{16} are truth-functions of p, any of other 12 is not a truth-function formed by p only.

Perhaps some one may think that to completely define non-truth-valued function in this sense, we have to use every truth-function formed by one or several variables and classical connectives to uniquely correspond to each non-truth-valued function which formed by each different non-truth-functional operator and one of the variables. For example, $\diamond_2^q p \overset{def}{=\!=} p \vee q$, $\diamond_2^r p \overset{def}{=\!=} p \vee r$. Nevertheless, it is obvious that this kind of definitions is not according with natural language. For example, even if we believe the "ought to" is a non truth-functional connective, we never think there are two different connectives in "I ought to eat food" and "I ought to drink alcohol". The recurrence of "ought to" is just the same!

2.2. *A Formal Definition of Non-Truth-Function*

In CP, basic symbols are:

(1-1)p, $p_1, p_2, \ldots p_n, \ldots$; as propositional variables form an order Pt which is a countable infinite set, $n \in \{N\}$

(1-2) $\neg,\vee$

(1-3) (,)

And there are some auxiliary symbols in CP:

(2-1) A, B, C, A_1, B_1, C_1, as the formula set A_t;

(2-2) For any of 16 truth-functions binary connectives d_i, i $= 1, 2, 316$. For two any of 16 truth-functions binary connectives, they are d_i, d_j;

(2-3) for any two formulas in At, they are X, Y;

STRF completely reserves all classical propositional connectives and their interpretation, propositional variables, classical axioms and theorems, as well as other rules in the classical propositional logic system CP. The only difference is that it adds a new sentential operator H that is the simplified form of H_7 in Table 2 which can be entirely defined by other classical operators in CP. Therefore, the new system SCP is, in fact, still the CP system.

Definition 1: Basic unary operate H

$Hp \overset{def}{=\!=} p_1.$

Table 1. Basic Classical Bivalent Truth-Functions and Binary Connectives

p	p_1	D_1	D_2	D_3	D_4	D_5	D_6	D_7	D_8	D_9	D_{10}	D_{11}	D_{12}	D_{13}	D_{14}	D_{15}	D_{16}
1	1	1	1	1	1	0	1	1	1	0	0	0	1	0	0	0	0
1	0	1	1	1	0	1	1	0	0	1	1	0	0	1	0	0	0
0	1	1	1	0	1	1	0	1	0	1	0	1	0	0	1	0	0
0	0	1	0	1	1	1	0	0	1	0	1	1	0	0	0	1	0
Connectives		d_1	d_2	d_3	d_4	d_5	d_6	d_7	d_8	d_9	d_{10}	d_{11}	d_{12}	d_{13}	d_{14}	d_{15}	d_{16}

Table 2. Truth Table of Narrow Set of Basic Unary Operators

p	p_1	D_1	D_2	D_3	D_4	D_5	D_6	D_7	D_8	D_9	D_{10}	D_{11}	D_{12}	D_{13}	D_{14}	D_{15}	D_{16}
1	1	1	1	1	1	0	1	1	1	0	0	0	1	0	0	0	0
1	0	1	1	1	0	1	1	0	0	1	1	0	0	1	0	0	0
0	1	1	1	0	1	1	0	1	0	1	0	1	0	0	1	0	0
0	0	1	0	1	1	1	0	0	1	0	1	1	0	0	0	1	0
$p =$ H_6p	$p_1 = Hp$ H_7p	$p\vee$ $\neg p$	$p\vee$ H_p	$p\vee$ $\neg H_p$	$\neg p\vee$ H_p	$\neg p\vee$ $\neg H_p$	p	H_p	$(p \vee H_p)\wedge$ $(\neg p \vee \neg H_p)$	$(\neg p \vee H_p)\wedge$ $(p \vee \neg H_p)$	$\neg H_p$	$\neg p$	$p\wedge$ H_p	$p\wedge$ $\neg H_p$	$\neg p$ $\wedge H_p$	$\neg p\wedge$ $\neg H_p$	$p\wedge$ $\neg p$
Connectives		d_1	d_2	d_3	d_4	d_5	d_6	d_7	d_8	d_9	d_{10}	d_{11}	d_{12}	d_{13}	d_{14}	d_{15}	d_{16}

C(Corollary)1-1: other 15 unary operators can be recursively definite by set $\{\neg, \vee, H\}$ in Table 2

C1-2: $\underbrace{HH\ldots H}_{n}{}_p \overset{def}{=\!=} p_n,\ n \in \{N\}$.

C1-3: the set of 6 symbols {¬, ∨, H, p, (,)} can define any proposition and any formulas in SCP, since every variable can be defined by H and p, and any formula is formed by variables in Pt and set $\{\neg, \vee\}$.

For convenience, we use several auxiliary symbols in SCP:

(2-5) $\blacksquare_1 p, \blacksquare_2 p, \ldots \blacksquare_m p$ as general non-truth-function of p, and each of them can be equivalent transformed into a composed truth-valued function, and vice versa. And each of $\blacksquare_{1A}, \blacksquare_{1B}, \blacksquare_{2A}, \blacksquare_{2B},$, can be equivalent transformed both into a composed truth-valued function and into a general non-truth-function of p. For example, $\mathrm{H}_i\mathrm{A} = \mathrm{d}_i(\mathrm{A}, \mathrm{A}_1)$, $\mathrm{H}_i\mathrm{B} = \mathrm{d}_i(\mathrm{B}, \mathrm{B}_1)$,

$\blacksquare_p$ as any general non-truth-function of p in SCP.

We regard the truth-valued functions as a special kind of non-truth-valued functions.

Definition 2: Semantic representation

STRP maintains the semantics of CP unchanged; only the rows of the truth table are equivalent to rewrite as a single row truth table in which "/" represents the first dichotomy between true and not true (or false), "/ /" represents the second dichotomy, and so on.

For example, if the truth-value of p assigned to "1/0", then the truth-value of H_2p and the truth-value of H_2H_2p assigned to "1/1//0" and "1/1//1///0". when $H_2p \overset{def}{=\!=} p \wedge p_1$, its truth-value semantics is: "1//0/0".

C2-1: Negation: The negation of any formula is negation of each separate truth-value.

For example, the semantic formula of A is: 0/1//0///1, so the negation of A is: 1/0//1///0.

C2-2: Disjunction: For any two formulas A and B, $A \vee B$ is that the disjunction for every partition between $A \vee B$.

E.g. if A = 0/1//0///1, B = 1//0/1//0,then $A \vee B = 0/1//0///1 \vee 1//0/1//0 = (0 \vee 1//0)/(1//0///1 \vee 1//0) = ((0 \vee 1)//(0 \vee 0))/((1 \vee 1)//(0///1 \vee 0)) = ((0 \vee 1)//(0 \vee 0))/((1 \vee 1)//((0 \vee 0)///(1 \vee 0))) = 1//0/1//0///1$.

2.3. *Important properties of classical propositional logic revealed by STRP*

SCP is just another sight of CP, but SCP will form several important derived rules that are not easy to detect in the original CP system.

Rule 1: Differentiating operator with superposition rule

For example, if $\blacksquare p = Hp = p_1$, then $\blacksquare\blacksquare p = \blacksquare p_1 = p_2 \neq p_1$

Rule 2: Non-completely substitutability rule

For any non-truth-function $\blacksquare A$ replaces $\blacksquare B$, in general, according to the K-2 rather than the K-1:

K-1: $if\blacksquare A = d(A, A_1), then\blacksquare B = d(B, A_1)$;
K-2: $if\blacksquare A = d(A, A_1), then\blacksquare B = d(B, B_1)$.
In fact, when ■ is reduced to be a truth-valued function operator, then K-2 is reduced to K-1, or if the last variables shown in A and in B are the same, then K-2 is reduced to K-1.

2.4. *Explanation*

We believe that the problems of truth-functional logic should be resolved by virtue of improving truth-functional logic itself. We are familiar with the certainty of "the uncertainty" in quantum mechanics. Quantum mechanics cannot precisely foresee the predictive results of single quantum measurement but it can foresee the probability. However, the probability of the results of a single measurement that quantum mechanics can foresee is completely definite. This is also reflected in a pair of conjugate variables cannot be precisely measured with certainty at the same time, but they have the uncertainty relation that is for sure. Further, in two identical particles assemble, despite if what we could not definite this or that one of the two particles by a single quantum measurement, but this is definite that such a single measurement can be measured one or another of two.

Still take H_2p in the previous section as an example. First of all, when p is to definite truth value 0, the true value of H_2p is not definite either 1 or 0 but has uncertainty of truth value 1/0 (we call it can be true can be false but not be both true and false at the same time); but then, that truth value of H_2p is 1/0 is completely definite.

Obviously, this definition H_2p (and considering Hp, and so on) does not have a stacked and substitution of contradictions, but necessary to produce two important properties of Rule1and Rule2 in Section 2.3. That is to say, it should be able to find a way to definite non-truth-valued function which only by CP system of classical truth-function closely, consistently, compactly, and completely at the same time. That once regarded as puzzling "non-classical" characteristics presented in natural language or in scientific language turn out to be "classical" ones in fact.

Someone maybe think about that Pt in CP is not a fixed order, for example, if the order of these three variables is p, q, r, then HHp=Hq=r; If the order of these three variables is p, r, q, then H'H'p=H'r=q. I think that, no matter what the order is, r is r keeps no change. Though in the first order, $HHp \overset{def}{=\!=} r$; in the second order, $H'p \overset{def}{=\!=} r$ it is definite in each order r can only be defined a non-truth-valued function. So we get HH=def H' only.

3. Interpretation

3.1. *The failure of equivalent substitution principle*

According to STRF, no matter what p_1 exactly is in a non-truth-valued function of p that is $H_2p = p \vee p_1$: if p = 1, then $H_2p = 1$; if p = 0, then $H_2p = 1/0$. So, we can get a recursive definition for H_2H_2p: if $H_2p = 1$, then $H_2H_2p = H_2(H_2p) = 1$; if $H_2p = 0$, then $H2H_2p = H_2(H_2p) = 1/0$. In addition, we get this definition: if p = 1, then $H_2H_2p = 1$; if p = 0, then $H_2H_2p = 1/1//0$.

We order two formulas, A = p, $B = H_2$p. Obviously, A and B are two different formulas. But when p = 0 and $H_2p = 0$, A = p = 0; $B = H_2A = 0$, $H_2B = H_2H_2p = 1/0$. This shows that the result can be satisfactory: when A = B, $H_2A \neq H_2B$. To accommodate the result, according to STRF, H_2H_2p in syntax has to be $p \vee p_1 \vee p_2$, but can not be $p \vee p_1 \vee p_1$.

To sum up, we get "a failure of equivalent substitution principle" just in classical logic system CP only: when p = 0, $p_1 = 0$, and $p_2 = 1$; A = B = 0, but $H_2A = H_2p = p \vee p_1 = 0$, $H_2B = p \vee p_1 \vee p_2 = p_2 = 1$. That is to say, there is a group of truth assignment, A and B are equivalent, but when we use B to substitute A of H_2A, then H_2A and H_2B are not equivalent.

Of course, "a failure of equivalent substitution principle" here is truly "non-classical", for "H_2" is not classical logic connective but a non-truth-valued functional operate, but we needn't intensional logic or any other non-classical logic to interpretation it.

In fact,"the failures of equivalent substitution principle" both in artificial language and natural language are more complicated, someone may think the example above is ad hoc or oversimplifying. We are going to reduce some "non-classical" modal characteristics next.

3.2. *Preliminary interpretation of modern modal logic*

Modal logics have become a huge non-classical (or non-standard) logical cluster beyond classical logics. From a philosophical point of view, it can be viewed as a kind of extended logic which adds some non-truth-functional logical operators and thus some logical axioms or rules without revising any classical logical axioms or rules (Susan Haack 1978). For the sake of simplicity, modal logics can also be considered as belonging to 'the class of logics which differ from standard logic for being linguistics or axiomatic extensions of it' (W. Carnielli and C. Pizzi 2008, chap. 1). Speaking in narrow sense, modal logics are mainly studying on reasoning that involves the use of the expressions 'necessarily' and 'possibly', but as necessity and possibility have kinships with many other philosophically important expressions, modal logics in broad sense cover a whole family of logics with similar rules and a variety of different operators, such as tense logic, deontic logic, epistemic logic etc (Garson 2006, p. 1–2). As we know, many different formal systems have been constructed for modal logics. The well-known modal logical systems are K, D, T, S4, B and S5, and all of these systems can be thought of as intending to provide an approximating adequate account of necessity (signified by operator '$\Box$') and possibility (signified by operator '$\diamond$'). Thanks to the works of Kripke and Hintikka and many other philosophers, possible worlds semantics are created to correspond to modal logical systems and a mature model theory are therefore established (Copeland (2002, 2006) and Korte, *et al.* (2009) give a detailed introduction to the development of possible worlds semantics). Although modal logics suffered from some critiques and suspicions about its syntax and semantics (see Quine 1947; Peregrin 1993; McCarthy 1997; Rescher 1999; Divers 2006), it is still very successful and useful. Especially, the possible worlds semantics has been used to adapt to many other non-classical logics. So far, philosophers and logicians' attention have been paying to specific logical problems within modal logic and some relevant philosophical issues and its applications (Blackburn, *et al.* 2002; 2007, Hughes and Gresswell 1996; Cocchiarella and Freund 2008). In addition, some philosophers and logicians have touched upon the problem about the relationship between modal logics and classical logics/standard logics. For example, some philosophers treated modal logics as fragments of standard logics and under this view modal logics inherit their semantics from the standard semantics of classical first or even higher-order predicate logic but restricting expressive power by using operators instead of explicit quantification (see Blackburn, *et al.* 2007, preface p. xi.). In other words, "modal logic is better viewed as a methodology for tapping into a core theme in standard logic" (ibid, p. xi). The link between standard logic and modal logics, as W. Carnielli and C. Pizzi (2008, chapter 1: modal logic and standard logic) said, 'is so strong that the dominant attitude shared by contemporary logicians, until the first decades of the 20th century, was to deny the legitimacy of an independent modal logic', and thus 'modal logics are not essentially rivals of standard logic, but are linguistic and axiomatic enrichments of it'. We hold this idea that modal statements could be paraphrased into the language of classical logic. Though most of scholars consider the modal formula with possible worlds semantics corresponding first-order standard logic formula, we think modal propositional logic corresponding classical proposition logic at first, for normally, first-order standard logic can be reduced to classical proposition logic in an individual domain with definite and finite. We have analyzed the certainty of uncertainty last section, so we think that the reduction of modal propositional logic in two-truth-valued and finite domain of individual to classical proposition logic at this sub-section from a new perspective STRF. The previous failure of such reduction is because a correct method has not been found.

To make things simple and precise, this section consider atomic modal proposition at first. Of course, "□"is operate "necessary", but from pure formal view, "□" is a symbol called "box" only. We needn't to know the detail means of the "□" in nature language and the long history of "necessary" in philosophy of logic. It is well-known that □p is a bivalent unary non-truth-function of p, according to STRF, any basic bivalent unary non-truth-function may always be equivalent transformed to a truth-valued function. So we can use this equivalent transformation to consider every □p in any modal schema. Take T schema $\vdash \Box p \to p$ for an example. We can find that:

(a) "□" in T schema represents a set of (but not a) truth-functions.
(b) T schema actually represents a set of theorems of classical propositional logic.

(c) Any truth-functions represented by $\Box$ which make T schema valid is iif this form: $p \wedge \blacksquare p_k$. k is the last valuable shown in $\blacksquare p_k$. when k = 0, then $\Box p$ represents p only; when k = 1, then $\Box p$ represents four truth-functions: $p \wedge p$, $p \wedge (\neg p)$, $p \wedge (p \vee \neg p)$, $p \wedge (p \wedge \neg p)$; when k = 2, then $\Box p$ represents sixteen truth-functions which are formed by p and p_1; and so on. That is to say, in T schema, $\Box p \overset{def}{=\!=} p \wedge \blacksquare p$, $\blacksquare p$ is "any non-truth-valued function of p".
(d) $\Box p \overset{def}{=\!=} \forall x(Fx \to (x \to p))$, in first order logic. Fx: x is a truth-valued function.
(e) In nature words, "necessary p is true" in T schema is "it is always false that any premise infers negation of p", for $p \wedge \blacksquare p = \neg(\neg p \wedge \neg \blacksquare p) = \neg(\neg \blacksquare p \vee \neg p) = \neg(\neg \blacksquare p \to \neg p)$.

So, we obtain that in T schema $\vdash \Box p \to p$, $\Box p$ represents a set of non-truth-functions of p which are equivalent to truth-functions one by one. And the set of truth-functions is a sub-set of the whole set of truth-functions in classical propositional logic system SCP, in which each such truth-function makes $\vdash \Box p \to p$ valid.

According to "possible world semantics" that $\Box p$ is true in actual world in T system represents p is true in all of accessible possible world which are reflective to actually world. Fortunately, any truth-function formed as $p \wedge \blacksquare p$, such as p, $p \wedge p_1$, $p \wedge \neg p_1$, $p \wedge \neg p$, etc, which lets T schema valid is and only is the truth-function in CP which is Reflective to p in the sense of the set theory.

To make things more simple and precise, we calculate ten common sachems about modal formula with no superposition, we call this form is SCP-B, and find some more interesting result at Table 3.

$\Box A = f(A, A_1)$; $\Box B = f(B, B_1))$ $\Box\Box p = f(f(A, A_1), A_2)$.
K:$\vdash \Box(A \to B) \to (\Box A \to \Box B)$; D:$\vdash \Box A \to \Diamond A$; T(reflective):$\vdash \Box A \to A$; Tr:$\Box AA$;
4 (transitive):$\vdash \Box A \to \Box\Box A$; B(symmetry):$\vdash A \to \Box\Diamond A$; E(Euclidean):$\vdash \Diamond A \to \Box \Diamond A$;
M:$\vdash \Box \Diamond A \to \Diamond\Box A$; O:$\vdash \Box(\Box A \to A)$; V:$\vdash \Box A$;

Checking row by row, each of 9 valid functions in D row is Seriality to A in the sense of the set theory, each of 4 valid functions in T row is Reflective to A, each of 5 valid functions in 4 row is Transitive to A, each of 4 valid functions in E row is Euclidean to A, each of 5 valid functions in B row is Symmetry to A, and so on in other rows.

Let us check among different rows. The equivalent relation among validities of various axioms is also complied with almost every classical result of modern modal logic (see Li Xiao-Wu, 2006, p. 112–113).

(1) Reflexivity $\to$ seriality.
(2) Symmetry + transitivity $\to$ Euclidean-ness.
(3) Omit (because we have not understood it).
(4) Reflexivity + Euclidean-ness $\to$ symmetry.

Table 3. Basic Classical Bivalent Truth-Functions and Binary Connectives

			Seriality	Reflective		Transitive	Symmetry	Euclidean			
Connectives/axioms	C	K	D	T	TR	4	B	E	M	O	V
$D_1 = \Box A = A \vee \neg A$	v	v				v	v	v		v	v
$D_2 = \Box A = A \vee A_1$	v					v		v			
$D_3 = \Box A = A \vee A_1$	v					v		v			
$D_4 = \Box A = \neg A \vee A_1$	v	v					v				
$D_5 = \Box A = \neg A \vee = \neg A_1$	v	v					v				
$D_6 = \Box A = A$	v	v	v	v	v	v	v	v	v	v	
$D_7 = \Box A = A_1$	v										
$D_8 = \Box A = (\neg A \vee A_1)$ $\wedge(A \veebar A_1)$	v	v	v								
$D_9 = \Box A = (\neg A \vee \neg A_1)$ $\wedge(A \veebar A_1)$	v	v	v								
$D_{10} = \Box A = \neg A_1$	v										
$D_{11} = \Box A = \neg A$	v	v	v				v		v		
$D_{12} = \Box A = A \wedge A_1$	v	v	v	v							
$D_{13} = \Box A = A \wedge \neg A_1$	v	v	v	v							
$D_{14} = \Box A = \neg A \wedge \neg A_1$	v	v	v						v		
$D_{15} = \Box A = \neg A \wedge A_1$	v	v	v						v		
$D_{16} = \Box A = A \wedge \neg A$	v	v	v	v		v			v		

(5) Symmetry + Euclidean-ness → transitivity.
(6) Symmetry + transitivity = symmetry + euclidean-ness.
(7) Symmetry + transitivity + seriality = reflexivity + euclidean-ness.
(8) Reflexivity + symmetry + transitivity = reflexivity + euclidean-ness.

Obviously, K axiom is not valid for all of 16 truth functions, at least, invalid for D_2 (or D_3). Namely, $\Box A = H_2 p_1 = p_1 \vee p_2$ and $\Box B = H_2 p = p \vee p_1$ in D_2, then K axiom is not valid when p, p_1 and p_2, are designated 0, 0, 1 respectively. For $\Box(A \to B) \to (\Box A \to \Box B) = H_2(A \to B) \to (H_2 A \to H_2 B) = H_2(p_1 \to p) \to (p_1 \vee p_2 \to p \vee p_1) = 1 \vee p_k \to (1 \to 0) = 0$. It is very interesting that K axiom is valid for all of 16 truth functions in K-1 but not K-2 (see Rule2 in 2.3). Since K axiom is valid for all of models in possible world semantics, so we can believe that possible world semantics is proper to K-1, while modal syntax is proper to K-2. If STRF can completely account for any modal non-truth-function as a set of truth-functions in classical logic, then we can say that the research results having been with syntax in modal propositional logic are almost correct, but not for all of the results with possible world semantics. Moreover, it may explain why, within possible world semantics, only most of but not every modal logic axiom(s) corresponds to a first-order predicate formula one by one, and vise visa.

This article has not yet completely reduced possible world semantics to classical semantics, but whatever kind of semantics we adopt, if STRF holds, modern modal logic can be treated as fragment of classical propositional logic which makes systematic classification studies on classical truth-functions. Syntactically, $\Box p$ in any axiomatic schemas of modal propositional logic represents a set of classical many-variable truth-function formed by p in which p is one of the variables which make the axiomatic sachems valid. Any modal axiomatic schema is and is only a set of classical propositional logical theorems. The possible world semantics more or less reflexts the unification of syntax and classical semantics: a possible word w is (or maps) a classical truth-function and a set W of possible worlds represents such a set of truth-functions. The corresponding relation between possible worlds is a common set property in terms of these truth-functions. W, R and classical valuation V in the form of K-2 should constitute a frame. That any axiomatic schema is syntactically valid in a frame means that if we substitute the every truth-function belongings to W into each "□"of the axiomatic schema in light of K-2 rules, a set of classical propositional logical theorems or a first order predicate logical theorem will be formed.

More over, some logicians which are the inventers of possible world semantics may not efficiently think about the different between these two things of "true": "a proposition p is true" in some condition that p is in a possible world w; "a proposition p in some condition that is a possible world w" is true. Obviously, the former is about proposition p's truth, but the later is about another proposition's truth, which is not p. Formally, no matter what the possible world condition of a proposition p which turns to be true or false is, the truth of $\Box p$ is always the truth only.

In fact, any set Ω can be divided into two subsets ε of Ω and non-ε of Ω. That is to say, for modal semantics of possible worlds, the most simple analysis method is that is, at first, divided the collection of all possible worlds into the real world ε and not the real world $\neg\varepsilon$. And we can always find another criteria β for the classification which is independent of ε, then divided $\neg\varepsilon$ into two sub-sub collections $\neg\varepsilon$ of β and $\neg\varepsilon$ of $\neg\beta$. Of cause, $\neg\varepsilon \cap \varepsilon$, $\varepsilon \cup \beta$, β, etc apparently are subsets of possible worlds set Ω. This is said that after two times of complete dichotomy, there are only 16 subsets of possible worlds. This shows that, although the "world" is very special importance in natural language, there is no any priority in the formal language.

Of course, the model in this sub-section about modal propositional logic is only a very simple one, we are going to do next is to restore the modal characteristics more in this definition method. In fact, the first modal characteristic is non-completely rule substitutability having revealed by our non-truth-functional definition in sub-section 2.3.

3.3. *An interpretation to deontic proposition*

Since it has almost inevitably been deontic paradox when deontic operator "obligatory" or "ought to" is directly interpreted into modal operate "necessary"in some system of modal propositional logic (see von Wright 1951, 1953, 1954, 1956, 1963, 1968, 1971; Anderson A. N. 1956, 1967, 1962b; Williams, B. 1965, 1980, 1996; Quine, W. 1962, 1976; Jackson, F. 1986, 1988; Goble, L. 2003, 2005; McNamara, P. 1996a, 1996b, 2004a, 2004b, 2006; Belnap, N, M. Perloff, and M. Xu 2001; Gabbay, M. 2002, 2006; Vranas, P. 2008, 2010; Moretti, A. 2009a, 2009b) we try to interpret deontic operator as one superposition of non-truth-valued functions in SCP, for any non-truth-valued function has to be formed by one or a set of superposition of 16 basic non-truth-valued functions according to our analysis in last section. Many humanistic philosophers often unarguably talk of "value judgment" or "moral judgment" in the same manuscript but in two different ways:

(1) It is ought to be that p.
(2) It is Good that p.

Therefore, we have three "operators" in which we can use to construct most basic deontic logic axiom: "ought to", "necessary" and "Good". So it is very easy (even for an elementary school pupil) that "ought to" means "necessarily Good", and we try to define basic atomic deontic propositions like this:

(7-1) $\bigcirc p \overset{def}{=\!=} \Box D(p)$

(8-1) $P_p \overset{def}{=\!=} \diamond D(p)$.

"D(p)" represents the corresponding actual proposition: p is deontic (or "p is Good"). Such defined deontic logical operators is consistent with basic axioms of unary operator logic in STRF. But we soon recognize that such definition is not the same as "$\bigcirc$" and "P" in present popular deontic logic, though it can formally reduce deontic logic into SCP perfectly. In present popular deontic logic, "should be", "ought to be" and "obligate" are usually represented as "$\bigcirc$"; "permit", "no prohibition" and "not compel not" are represented as "P", and $\vdash \bigcirc p = \neg P\neg(p)$. But in our last definition, $\bigcirc p = \neg P\neg(p)$ is just a satisfier while $\bigcirc D(p) = \neg PD\neg(p)$ is valid. Moreover, in natural language, it is obvious that "prohibition" means "not permit", and "obligation" or "compel" means "prohibit not".

To resolve this problem, let's return to humanistic context. We can see that "deontic" in those contexts mean "moral and obligatory". Some theorists have investigated into "moral logic" which is different from "obligatory logic", and somebody think that both deontic morality and obligation belong to philosophy of value. So let's make another metaphysical assumption: deontic logic includes and only includes logic of morality and logic of obligation. The former is the logic that investigates a content of value, namely, the logic about "Good". The latter is the logic of investigating in the other forms of value, namely, logic about "norm". Moral Good has only the feature of "persuasiveness", while norm of obligation has the characteristic of "coercion".

Therefore, although the symbol "P" (permission) which appears in both logic of morality and logic of obligation are as $P_p \overset{def}{=\!=} \diamond G(p)$ and $P_p \overset{def}{=\!=} \diamond D(p)$ which in fact are same both in logical syntax and in semantics, "$\bigcirc$" as "ought to" and "$\bigcirc$" as "obligated" are different represented by "$\bigcirc_u$" and "$\bigcirc_b$". In logic of morality, $\bigcirc_u p$ means "it ought to be that p" or "it is necessary that p is Good", while P_p means "it is permission that p" or "it is possible that p is Good". In logic of obligation or norm, $\bigcirc_b p$ means "p is obligation", but it does not mean that "it is necessary that p is normative". P_p means that "p is permission" and "it is possible that p is obligation". F_p means that "p is forbidden", and indeed "p is not permitted" means "p is forbidden", that is, "not-p is coerced".

What we need to take notice of is that in deontic logic to say "it is forbidden doing something" is to say "it is impossible to be normative to do something", because $\bigcirc_b \neg p = Fp = \neg Pp = \neg \diamond D(p)$. And to say "it is coerced to do something" or "it is obligation to do something" is to say "**it is not possible that not to do something is to be normative**", because $\bigcirc_b p = F(\neg p) = \neg P \neg p = \neg \diamond D(\neg p)$. But at the same time, we should note that $\bigcirc_u p = \Box G(p) = \neg \diamond \neg G(p)$ in logic of morality, which means "**it is not possible that to do something is not to be moral**". Then we obviously have:

Both in logic of morality and in logic of normality: $P_p = \diamond G(p)$, $P_p = \diamond D(p)$; but $\neg P \neg p = \neg \diamond G(\neg p) \neg \diamond \neg G(p) = \Box G(p)$, $\neg P \neg p = \neg \diamond D(\neg p) \neg \diamond \neg D(p) = \Box D(p)$. $\vdash G(p) \to P_p$, $\vdash D(p) \to P_p$; but $p \to P_p$ are not universally valid.

In logic of morality, $\bigcirc_u p = \Box G(p) = \neg \diamond \neg G(p) \neq \neg P \neg p$; but $\bigcirc_u p \to Pp$ is not universally valid.

In logic of normality, $\bigcirc_b p = F(\neg p) = \neg P \neg p = \neg \diamond D(\neg p) \neq \Box D(p)$; but $\vdash \bigcirc_b p \to P_p$.

In what follows, we use the basic definition of deontic logic above to solve the problem (T). Let "I eat food" as p, "I drink alcohol" as p_1; "I ought to eat food" as $\bigcirc_u p$, "I ought to drink alcohol" as $\bigcirc_u p_1$. Therefore, $\bigcirc_u p =$

$\Box G(p)$, $\bigcirc_u p_1 = \Box G(p_1)$, even thought we don't know exactly what kind of the necessary "□" is.

To make things simple, we stipulate traditional mean of "necessary": $\vdash \Box A \to A$.

Intuitionally, the mean of $\Box A$ here in naturally language is "anyway, A", or more convinced "it is always false that any premise having arbitrary selected infers negation of A". According to STRF, we recursively get $\Box G(p) = G(p) \wedge G_x(p)$, $\Box G(p_1) = G(p_1) \wedge G_y(p_1)$. $G(p) = p_k G(p_1) = p_{k+1}$ and so on. Then there is a set of truth assignments: p and p_1 are both 1, G(p) is 1 and $Gx(p)$ is 1, but $G(p_1)$ is 0, so that $\bigcirc_u p$ is 1 but $\bigcirc_u p_1$ is 0. In natural language, that is: "any premise does not infer I eat food is not Good"; "at lest one premise does infer I drink alcohol is not Good." In other words, when p=q, $\bigcirc_u p \neq \bigcirc_u q$. Thus, we solve the formal problem (T) only by virtue of axioms, theorems, and rules of classical logic about the truth-function.

3.4. *An explanation to some many-valued logics*

In classical logic, the "split" of true and false which is not-true in truth-value domain is a complete dichotomyso there is impossible to be a truly third truth-value. For our sake, we divide the so-called "the third truth-value" in three-valued logics into several categories as following:

3.4.1. *No truth-value definition*

Some philosophers viewed "the third truth-value" in three-valued logics as "no truth-value at all", and this kind of three-valued logic can seemly be used to interpret a sentence with non-referential terms. As D. J. Shoesmith and T. J. Smiley said, a propositional calculus is many-valued if and only if it is normally constituted and it has the cancellation property.[2] According to Smiley, when a third logical value is assigned to a well-formed formula, it doesn't mean that the formula has uncertain truth-value, but means that it has no truth-value at all. We will not focus on this type of "the third truth-value" now, because obviously it is not in the scope of our STRF, though we had found a set of definite no truth-values in our formal research.[3]

3.4.2. *Truth-value "un-decidable"*

Some logicians (especially mathematicians) have a different understanding to the third logical value. An orthodox view usually attributed to Frege is that the meaning of a statement is given by the conditions under which it is true, its truth conditions. And truth as commonly conceived is a relationship between language and an extra-linguistic reality. However, there is a problem concerning what is corresponding

[2]See D. J. Shoesmith and T. J. Smiley, 1971, Deducibility and Many-Valuedness, the Journal of Symbolic Logic 36 (4): 610–622. p. 611.

[3]See Xiao-long Wan "The first exploration to unitory operators: Whether can unitory operators be exhaused?", in Journal of Anhui University (social Science edn, in Chinese), 6th, 2011.

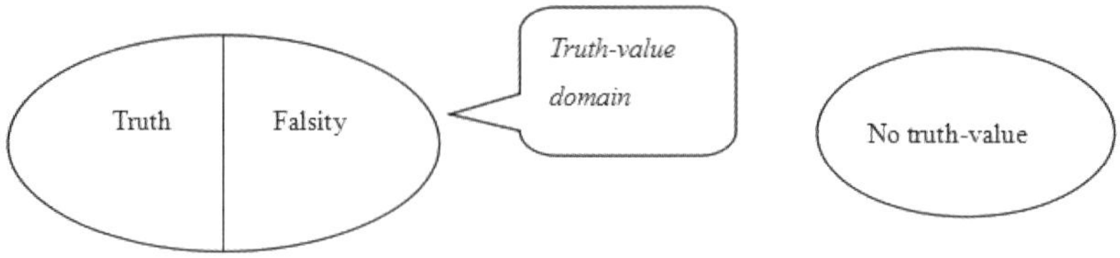

Graph 1

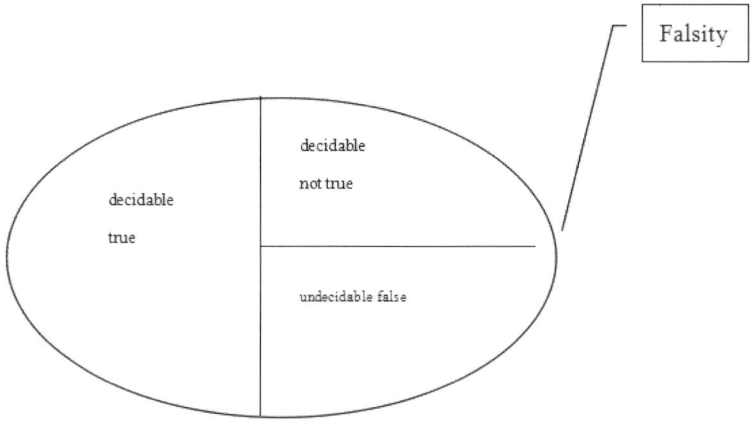

Graph 2

to the truth of a mathematical statement. The intuitionist hold that the meaning of a sentence is to be given by the conditions under which it is proved, its proof conditions, where a proof is a construction of a certain kind. In this case, we know that some famous mathematical conjecture whose status is currently undecided, such as the Twin Prime Conjecture, and Goldbach Conjecture. The mathematician S. C. Kleene also thought that many mathematical statements cannot be either true (proved) or false (not proved). He established a famous three-valued truth table in 1938 and 1952. A third truth-value is attributed to the un-decidable mathematical statement. As for us, this kind of third truth-value is confusing for since they regard truth-value true as a decidable true, then the second truth-value false which is not true have to be all of situations that is not decidable true which, of course, includes both decidable not true and undecidable not true, seeing Graph 2.

3.4.3. *Changing the definition of classical true or classical false*

In fact, many logicians inadvertently, changed the definition of classical true or classical false, when they introduced "the third truth-value". For example, that viewpoint (2) changed classical false. Specifically, the domain of false in classical sense is divided into two sub-domains of "false" and "neither true nor false" in many three-valued logical systems. Of course, it could be consistent, but obviously

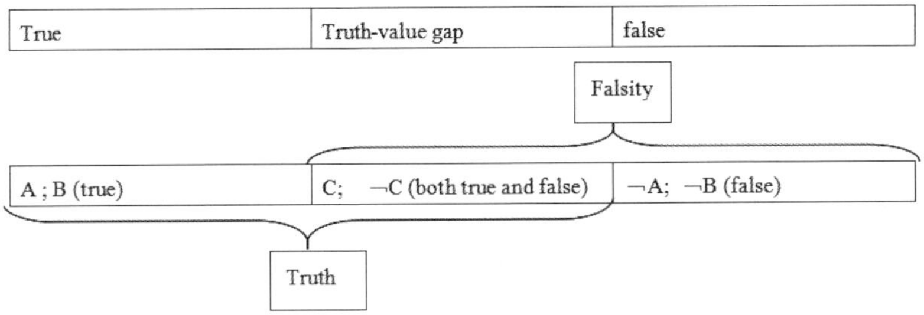

Graph 3

both "false" and "neither true nor false" are belong to classical false, it is not directly in the scope of STRF which just concentrate on the classical true and classical false.

3.4.4. *"The genuine third truth-value"*

This is the situation which needs serious discussion. For example, the quantum logic in which the logical domain of truth-value is the set of real numbers from 0 to 1, keeps the meaning of classical truth (1) and falsity (0) unchanged. To make things simpler, we simplify the situation as a three truth-valued one at this sub-section: "1" signifies "classical true", "0" signifying "classical false", and "1/2" signifying "the third truth-value" which means "neither true nor false" but nor no truth-value definition. There are two interpretations to "the genuine third truth-value". The first is that it is just a sub-domain of either classical truth or classical non-truth, which still resembles the situation (3), but it is not in the scope of STRF. The second is that it is neither "true only" nor "false only", and it also doesn't mean both true and false at the same time. We think that the so-called "genuine third truth-value" should be actually taken as a permutation of some combination of true and false. Considering traditional modal necessary proposition $\Box p$. When p = 1, then $\Box p = 0/1$. Logicians traditionally think that the truth-value of $\Box p$ here is uncertainty. But according to STRF, this uncertainty is another kind of certainty (determinacy), that is to say, it logically either can be true or can be false. And when someone incorrectly regards "0/1" as a single truth-value, it seems to be a third truth-value, but turns out to be a combination of true and false. This "genuine third truth-value" which can be reduced to classical two truth-values is exactly in the scope of STRF.

The nature of logic is dichotomy. With the formal complete dichotomy, classical logic clarifies and simplifies any complicated existence of transcendent/transcendental things, speculative philosophy and art, the image of literature and life, the advanced mathematical physics and other scientific theories, many minds or many worlds, and even every kind of logic theory itself, and keeps intactness and reliability at the same times. Even "non-can-be a complete dichotomy" and "can-be a complete dichotomy" are still referring to a complete dichotomy. The classical two-valued logic is actually a complete dichotomy of semantic formalization. In this sense, the truth-value of any proposition can and only can be generalized into (two-valued) true or not-true (i.e. false), which are the truth-values of the complete dichotomy. To make sure this point does not need to know what exactly the "true" is, because the relation between form "true" and "false" which is classical negative is independent of any specific philosophical theories of "truth" or "reference". Just as all things in the world can always be divided into x and not-x, there can't be a third case and it is unnecessary for us to know what is x. And based on the semantics of logical dichotomy, no matter what it represents with respect to truth-function or non-truth-function, the logical meaning of truth-value is always the same.

However, some scholars may doubt our theory: how could "indeterminacy" be a permutation of a combination of definite "true" and definite "false"? Modern truth-value theory can be traced back to Leibniz, his main point in today's popular words includes: a proposition is contingent true that is factually true; a proposition is ever true that is logically true. But I have to say: whether a proposition is contingent true or ever true, and whether a proposition is true in a factual world or in a possible world, when the proposition is true, it's true is of course the same logically true only, for proposition is a logical conception only. It seems that Leibniz and his followers confused with these two things that are the logically true of a proposition and the factually true of a sentence which represents the proposition. Even though we agree with, in the situation that a proposition and a sentence are corresponding one by one, that when the proposition is a logically true then the sentence which represent the proposition must be a factually true, and vise visa, we still can only agree with that the factually true of the sentence is logically equivalent to the logically true of the proposition but can not agree with that the factually true is the same as the logically true at this situation. The difference between the logically truth-value of a proposition and the factually truth-situation of a sentence corresponded is just indistinguishable when we only think about a single proposition and the single sentence corresponded, but it is distinguishable when we think about the relation between the logically truth-value relation between two propositions and the factually truth-situation relation between two sentences corresponded.

Since any logically truth-value of a proposition can only be either true or false, in the scope of our STRF, and the logically truth-value of a proposition and the factually truth-situation of a sentence corresponded is indistinguishable when we only think about a single proposition and the single sentence corresponded, any factually truth-situation of the sentence corresponded of course can only be either true or false, and can not be both true and false at the same time, and can not be any other factually truth-situation also in the scope of our STRF. That is to say, when we only think about a single proposition and the single sentence corresponded, both the proposition is impossibly a third truth-value "indeterminacy" and the single sentence corresponded is impossibly a third factually truth-situation "indeterminacy" in the scope of our STRF. But when we think about two propositions or two sentences corresponded at the same time, the result is subtle different. Now we think about the two sentences in the scope of our STRF at first, in any facture situation, the factually truth-situation of any one of the two sentences has to be either true or false only, regardless of whether they are future contingent factually situation or not. That is to say, in any factually situation, while the factually truth-situation of the first sentence has been a definitely one, the factually truth-situation of the second sentence has to be definitely either true or false only. Corresponding to the facture situation, both two propositions have definitely true-value either true or false only, regardless of whether they are truth-valued function or not. That is to say, corresponding to a factually situation of any two sentences, the two propositions is certain to be a group of truth assignments. Then we think about the logical situation of two propositions in the scope of our STRF. When the logic truth-value of the first proposition is determined, how about the logic truth-value of the second proposition in this logical situation? This question is obviously not mind that when the first truth-value is determined, then the second one must be determined either true or false as a definite truth assignment; but that when the truth-value of the first proposition is determined, then what possible adopt of logical truth-value of the second proposition. For example, when p is determined to be 0, then in this logical situation, H_2p is determined to be 1/0 that is not determined to be either 1 or 0 only. That is to say, in our ordinary thought custom, the logical situation and the facture situation are not always corresponding one by one. It is more obviously about problem (T) at the beginning of this article:

I always factually either ought to drink alcohol or ought not to drink alcohol, and correspondingly that I ought to drink alcohol is always either logically true or not true. Moreover, in any facture situation, regardless of whether I drink alcohol or not, I always factually either ought to drink alcohol or ought not to drink alcohol. However, regardless of I drink alcohol is either logically true or not true, that I ought to drink alcohol be logically possible true and logically possible not true, but can not be both true and not true at the same time.

From above, we know that, for any sentence which seems factually "indeterminacy", it is in fact that the proposition corresponded is logically "indeterminacy" with other proposition in which this is "determinacy" that the proposition always can be logically true and can be logically not true, but can not be both true and not true at the same time. Thus, many logicians are confused with two questions: The first question is that the truth-value of each of any two propositions could only be either true or false; the second question is that when one proposition takes a truth-value true (or false), the truth-value of another proposition can not be always logically true only (or false only), for the second one is a non-truth-valued function of the first one. There were many situations in which the so-called "the third truth-value" may be a permutation of combination of definite "true" or definite "false". For example, many logicians regarded that: the statement of "There will be a sea battle tomorrow" with regarding to "Today" be indeterminate in truth value. For our STRF, "Today" is an abbreviated proposition logically relevant to

Table 4. Lukasiewicz's L_3

f¬		**f∧**	**1**	**i**	**0**	**f∨**	**1**	**i**	**0**	**f⊃**	**1**	**i**	**0**
1	0	**1**	1	i	0	**1**	1	1	1	**1**	1	i	0
i	i	**i**	i	i	0	**i**	1	i	i	**i**	1	1	i
0	1	**0**	0	0	0	**0**	1	i	0	**0**	1	1	1

"There will be a sea battle tomorrow". "Today" and "There will be a sea battle tomorrow" are all bivalent. Therefore, the latter proposition is regarded as always a truth-function with n-variables, while the former expression "Today" is regarded as one variable among the n-variables. Even if we attribute an indefinite "third truth-value" to it, the third truth-value must be a permutation of the combination of definite "true" and definite "false", despite of our ignorance of whatever the permutation of truth-value combination is. We speculate that a vast majority of logician in favor of many-valued logic followed Lukasiewicz, who ignored the relationship between non-truth-function and truth-function. We all know that the construction of the first three valued logical system by Lukasiewicz stemmed from his future contingent accidence, or indeterminism.

The three-valued system of propositional logic owes its origin to certain inquiries I made into so-called "modal propositions" and the notions of possibility and necessity closely connected with them.[4]

Now we may explain why the law of contradiction and the law of excluded middle are seemingly failed in Lukasiewicz's three-valued logical system L_3. In the truth table of L3, there are three truth-values: 1, 0, and "i" in Table 4.

According to his three-valued truth table, Lukasiewicz thought that the classical negation of "the third logical value" i is i itself (neither true nor false). In other words, if a proposition A = i, then $\neg A = i$, so $A \wedge \neg A = i \neq 0$, $A \vee \neg A = i \neq 1$. Obviously the law of contradiction and the law of excluded middle are seemly failed in this system. As he said in On Three-Valued Logic (1920)

The laws of three-valued logic differ partly from those of two-valued logic. Some laws of Aristotelian logic are only "possible" in three-valued logic, for instance, the principle of the syllogism in the ordinary formulation: $(a < b)(b < c) < (a < c)$ *but the principle of the syllogism in the formulation:* $(a < b) < [(b < c) < (a < c)]$ *is true, the principle of contradiction aa'= 0, the principle of the excluded middle a+ a'= 1, etc. Some laws of two-valued logic are false in three-valued logic, among them the law: (a= a') = 0, since for a = 1/2 the sentence a= a' is true. This accounts for the fact that in three-valued logic there are no antinomies.*[5]

According to STRF, even we don't know whatever the truth-valued combination exactly is, we realize that the "indeterminacy" has to be a permutation of

[4] Lukasiewicz, 1930 Philosophical remarks on many-valued systems of propositional logic [J], in Jan Lukasiewicz selected works, edited by L. Borkowski, North-Holland Publishing Company, 1970, p. 153.

[5] Lukasiewicz, On Three-Valued Logic [J], in Jan Lukasiewicz Selected Works, edited by L. Borkowski, North-Holland Publishing Company-Amsterdam London, 1970, p. 88.

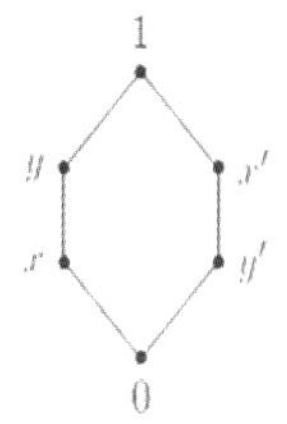

Graph 4

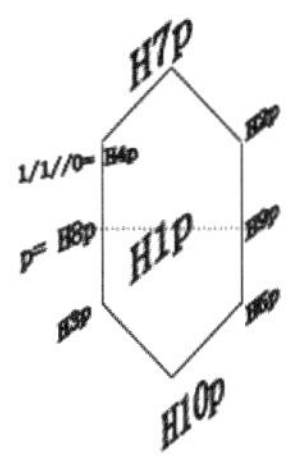

Graph 5

some combination of definite "true" and definite "false", and we can be sure that the classical negation of "indeterminacy" has to be another permutation of the same combination of definite "true" and definite "false" as the "indeterminacy" by virtue of the Rule 3-1 in section 2.3. For simplicity, we suppose "indeterminacy" is $0/1$, then the classical negation of "indeterminacy" is $\neg(0/1)$,

$\neg(0/1) = \neg 0/\neg 1 = 1/0$.

And it very easy to see that,

$0/1 \wedge 1/0 = 0 \wedge 1/1 \wedge 0 = 0/0 = 0$, $0/1 \vee 1/0 = 0 \vee 1/1 \vee 0 = 1/1 = 1$

That is to say, if Lukasiewicz keeps the domain of truth-value complete in his system, and then the law of contradiction and law of excluded middle still hold. In other words, if we only think about the degree of truth, there are in fact only three truth values in Lukasiewicz's system. But when we think of the distributions of the truth-degree of the third logical value, we discover that the thing has to follow its shadow. "One truth value, many distributions" is a very important feature in many-valued logic according to STRF.

3.5. *The minima classical explanation for Birkhoff-von Neumanns quantum logic*

The standard mathematical form of quantum mechanics and quantum logic are two aspects of the great contribution to quantum theory by von Neumann. After given a functional mathematical argument for quantum mechanics in 1932, von Neumann intended to provide an intuitive understanding for the abstract functional analysis in mathematical form in the article "the logic of quantum mechanics" (1936). That article is not only the beginning of the study on quantum logic — a non-classical logic which crosses philosophy and science, but also a start point of an important interpretation cluster of quantum mechanics constituted by other quantum logical approaches, such as many-valued quantum logic school by Reichenbach (1944), complementary dialectical logic school by Bohr and Rosenfeld, and modal logical interpretation school by van Fraassen. Moreover, quantum logic gives rise to a well-known debate between philosophers and logicians on whether quantum mechanics leads to abandon some basic principle of classical logic (for example, distribution law or bivalence). In the past of over 70 years, quantum logic has become a lattice theoretic research tradition of non-classical logic in trans-disciplines like physics, philosophy of science, logic and computer science, and some research schools have been formed. The recent papers on many fundamental fields of quantum logic are published in the international conference on "quantum logic" held in Universiteit Utrecht, Netherland in January 2008. For example, van Benthem and Sonja Smets (2011) attempt to combine popular Kripke's possible worlds semantics with non-classical information theoretic interpretation of quantum mechanics to form a dynamic logic, in which they conclude that von Neumann's non-distribution of quantum disjunction between is not necessary to hold. On the other hand, Deutsch and Chiara formulated a kind of "quantum computational logic" with $\sqrt{not}$ as primary parameter in recent two decades (Deutsch 1992; Cattaneo, Chiara, Giuntini, Leporini 2004). Based on STRF, we claims that quantum logic is not self-consistent when it applies lattice theory to represent non-distribution of quantum disjunction, and the same conclution but eisaer than van Benthem and Sonja Smets'.

Concisely speaking, standard quantum logic stems from such two forms: abstract orthogonal modular lattice quantum logic and specific Hilbert quantum logic. These two forms of quantum logic have similar semantics and both indicate that the non-distributive of quantum disjunction can be exactly represented by non-distributive lattice graph, but the latter form determines a stronger logic. Let's take a lattice graph (graph 4: the benzene ring) which is highly recognized in quantum logical texts as an example to explain the understanding of quantum logicians to quantum non-distributive (Engesser 2009).

Lattice theoretic quantum logicians think that the previous graph 4 is one of the best models which reflect various non-classical characteristics of *quantum logic*. In this graph, $0 < x < y < 1$ and $0 < y' < x' < 1$ reflect quantum partial order; $x \cup x' = 1$ reflects a quantum disjunctive relation with orthogonal complementary relation; and the two lines reflect the relation between two Hilbert sub-spaces with no direct logical relation. Moreover, there is an orthogonal relation between any two points in the same horizontal lines. And we can seemly always find such four quantum experimental propositions so that the following quantum disjunctive formula holds:

(I) the left-hand is $y \wedge (x \cup x') = y \wedge 1 = y$ and the right-hand is $(y \wedge x) \cup (y \wedge x') = x \cup 0 = x$. We call the fulfillment of condition 1 ($y \wedge x = x$) and condition 2 ($x \cup x' = 1$) and condition 3 ($y \wedge x' = 0$) at the same time.

We find that even if we substitute the sixteen classical binary connectives one by one for the "∪" in above formula, eight connectives have no distributive but no one can obtain the result of formula (I). This shows that if the above three conditions are all met, quantum non-distributive cannot be the feature of classical logical functions. Nonetheless, there are two possibilities for formula (I): non-classical logical feature and non-logical feature. According to lattice theory, for any two points in the same lattice line (represent two propositions), it is clear-cut that the truth-value of conjunction of them ("∧") takes the less truth value of the two conjuncts, and the truth-value of disjunction of them ("∪") takes the greater truth value of two disjuncts. But for any two points in different lattice line, the union of two propositions ("∪") is the least upper bound and the conjunction of two propositions ("∧") is the greatest lower bound, but the meaning of "the least" and "the greatest" is unclear. Specifically, $x \cup x' = 1$ and $y \wedge x' = 0$ in graph 4 first of all need further pure mathematical or logical explanation.

Since lattice is a kind of infinite but non-countable many-valued logic used in the quantum logic established by Birkhoff and von Neumann, we can also use the STRF to prove the non-allocation character is not possible to set up. The key point in any lattice as "benzene ring"[6] above is the logic relationship between any two points in the different lines but at the same horizon level.

For instance, If the truth value of any one of the four points x, y, x', y' in graph 4 is a rational number, then they are all rational. Then, the truth values of y and x' are that they have the same combination of definite "true" and definite "false" but the different permutations of the combination. Suppose the truth value of x, y, x', y' are 1/4, 3/4, 1/4 and 3/4, then they are in fact: 1//0/0, 1/1//0, 0//1/0 and 0//1/1 in STRF. And $y \wedge x' = 0$ are always not hold.

$$\begin{aligned} y \wedge x' &= 0//1/1 \wedge 1/1//0 \\ &= (0//1 \wedge 1)/(11//0) \\ &= ((01)//(11))/((11)//(10)) \\ &= 0//1/1//0 = 1/2 \\ &\neq 0 \end{aligned}$$

If the truth value of any one of the four points x, y, x', y' in graph 4 is a irrational, then they are all irrational. Suppose the truth value of x.y.x'.y' are: $1-\sqrt{2/2}$, $\sqrt{2/2}$, $\sqrt{2/2}$, and $1-\sqrt{2/2}$, then we can use Geometric methods: "——..........", "————......", "......————" and "..........——". It is very clear that $y \wedge x' \neq 0$.

In short, it has been clear why the non-distributive lattice is impossible: because the past theory can not clear explain the logical true-valued relationships of any two points in the same horizon between any lines in any lattice. According to STRF, they have the same value but different distribution of the value. Similarly, almost

[6]See Kurt Engesser *et al.*, eds. 2009, Handbook of Quantum logic and Quantum Structures, Elsevier, p. 9.

every many-truth-valued logic system has had no consideration of the distributions of the truth value.

Furthermore, it is not hard to recognize that the basic logical operator "$\sqrt{not}$" (if we make two times $\sqrt{not}$ to any logical formula, the result of circle operation is equivalent to the classical negation of original formula) of quantum computational logic, as a non-classical connective but not mystery. It can be a composition of basic unitary operators. For example, among the operator set including second superposition, the twice superposition of operator H8 in circular operation is also equivalent to classic negation of original formula, namely, $H_8 H_8 p = \neg p$.

Quantum information technology is one of the most promising mankind's new areas, but the mathematics of quantum mechanics and the logic of information theory paradigm require further analysis.

The core concept of quantum science and technology is quantum wave function (quantum state), the most important problem in quantum mechanics is a quantum measurement, the most typical feature in quantum physics is quantum probability, and the most critical point in quantum information is the quantum state of the correlation. However, in accordance with STRF, from the definition of quantum state and its measurement, quantum mechanics does not seem a logical self-consistent theory. Consider the simple case of a single quantum state of a microscopic particle, based on the measurement theory, before a measuring, it may be in a variety of 1-bit or 0 and any real number between two, but after the measuring, it is of 1/2 possibility either in 1 or in a 0 bit, and that we actually see is determined in 1 instead of 0 (or at 0 instead of 1). It means that the logical relationship between 1 and 0 is opposition before the measuring, but the logical relationship between 1 and 0 is the classic negative relation after the measuring. That is to say, the logical relation between truth-valued 1 and 0 is the classic negative relation and is not the classic negative relation in the same von Neumanns quantum logic system.

So, how to deal with that kind disharmony in quantum mechanical theory? One approach is thoroughly exterminating the conceptual development of quantum mechanics in every step, and reflecting the predictions of quantum mechanics with the observable result one by one, which we have to do in the future work. Another approach is to first believe that von Neumanns quantum logic is correct simplified model from his own standard mathematical form of quantum mechanics, and reflect on the model associated with the many-valued logic and modal logic substance. This is the method used in this paper.

We have two choices now. A weak point is the quantum logic is indeed an infinite set of real numbers of many-valued logic. In order to solve quantum difficult, we must extend our STRF which can only deal with set of rational numbers between 1 and 0 to a new theory NTRF which can deal with set of real values between 1 and 0. Another strong point is that there is impossible to have a self-consistent logic system which is set of real numbers of many-valued. Therefore, according to the second point, self-consistent quantum logic can only be many-valued with a set of rational numbers between 1 and 0 which must be the classic negative relation, and seeing now the quantum theory of it as a real number set is not right.

If quantum logic happens to be a many-valued logic with rational arrangement of infinity, then we not only can solve the quantum measurement problem, but also should be able to reproduce all observable quantum mechanics representations with STRF, and can distinguish "principle of equal probability indistinguishable", because equal probability is the same combination of values 1 and 0 with different permutations. However, for some by the export of quantum information "non-classical", but may produce different results.

Let's start with measurement problems began. For example Schrodinger cat, according to the mathematical theory of measurement, after measuring, the cat state is the superposition of 1/2 probability of live cat state and 1/2 probability of dead cat state, but we always see the cat either alive or dead, how to explain this? Decoherence theory can explain why we do not see the superposition state, that is able to explain difficult A: what we see is always a certain 1/2 probability realized (one-another); but it can not explain the difficult B: Why do we see this time to be this one of 1/2 probability realized rather than that one (this-that)?

The same probability does not mean the same possibility. According to STRF, $\diamond p$ means "p is possibility". We know $\diamond p$ semantics is "p is 1, $\diamond p$ is 1; p is 0, $\diamond p$ is 1/0" can represent a probability p is 1/2. But how about $\diamond' p$, when $\diamond' p$ semantics is "p is 1, $\diamond' p$ is 1; p is 0, $\diamond' p$ is 0/1"? Of course, $\diamond' p$ can represent a probability p is 1/2 also, but obviously $\diamond p = p \vee p_1 \neq \diamond' p = p \vee \neg \diamond p$. Obviously, these discussions depend on the results of a thoroughly study of modal logic.

The observable affections in quantum mechanics have a bottom line that is the quantum probability. Therefore, only at this level can quantum mechanics not be able to distinguish the different possibilities with the same probability. We believe that is the cause of quantum measurement problem remains unresolved. Similarly, there is not consciously default of "equal probability of undistinguished principle" in the reasoning of Bell's inequality and of quantum information "no-cloning principle" and of other well-known rules or principles.

So, from the basic definition of STRF, we can also see Einstein on quantum mechanics requires "a hidden variable" intuition is profound, although his "the hidden variable" with EPR argument is wrong. Neither as D. Bohm's "hidden variable theory" which plus a new term in Schrödinger equation, nor as J. Bell-type "hidden variable prove" which plus a new variable under the wave function, our "hidden variable method" remains both Schrödinger equation and wave function unchanged but regards that the original two variables-wave function can be equivalence to a function in the strict sense with more than two variables, or the wave function under Hamilton operator acted can be equivalently transformed into a function in the strict sense in which the original wave function as one of the variable.

And finally we face to quantum information. If after we check the all observed foresee in quantum mechanics, we still find quantum state can be represented by the set of rational numbers, we maybe use STRF to understand the deep nature of a non-locality correlation between some non-functions in the strict sense as a normal feature between some functions in the strict sense. And we can get a definition of information which is the difference of uncertainty between two things which form a equation, forming information is to make equivalent transformation directing to the larger uncertainty one. Fundamentally, quantum entanglement reveals forming an information is independent to the disseminating the information, which is the key difference from the classical information.

3.6. *A limited appearance of the formal nature of the dialectical logic*

Since Hegel's *Science of Logic* appeared in 1800's, it has been received the much enthusiasm which though it was perhaps the best understood as a version of what Kant had called "transcendental logic" in ontological claims, but there are a few contemporary logicians who regard it can be understood as a treatise in formal (or "general") logic. In China, there are several groups of scholars who have been trying hard to formalize the Dialectical Logic in the sense of Marx and Engels tradition. And since 1950's, there are several international schools applying non-classical logics to reveal the formal nature of dialectical logic. We just focus on two typical approaches: dialetheism and paraconsistent logics.

3.6.1. *The minimum consensus*

According to most of those Chinese scholars, the core of the Hegel's Dialectical Logic is the law of negation of negation. Generally, the law is expressed by thought process of the triad: "thesis, antithesis, and synthesis". Negation of negation includes the two-step (i.e., dialectical negation):

Step1: from the thesis to the antithesis: the negation of the thesis.

Step2: from the antithesis (i.e., the negation of the thesis) to synthesis.

According to Hegelian philosophers, a truly dialectical negation is a negation of negation. After the first negation, the antithesis is deny and beyond the thesis, because it is further understanding of the thesis from the opposite side. After the second negation, it seems to be back to the beginning of the thesis, but in fact, it achieves the synthesis which includes both the thesis and the antithesis. In the view of some Chinese logicians, dialectical negation and classical logic negation are different, negative concept (or proposition) and negated concept (or proposition) in dialectical negation process can be the same true as well as the same false (this long-term uncorrected thought had once been the view that the dialectical logic is neither complied with the law of contradiction nor law of excluded middle in classical logic). Hegel further believed that a whole process of the dialectical logic can not be end of a two-step dialectical negation process, but a spiral series ascend expanded by numerous such processes, in which the concept is self-motion until to the absolute spirit itself. However, in the view of some speculative philosophers, dialectical negation is inner self-negation which is the logic of content that can not be formalized.

In our opinion, when we study the dialectical logic, we should distinguish Hegel's words about the dialectical logic which he invented from the nature of his dialectical logic which Hegel did not understand precisely. Any idea or thought, as long as it can use language to express, can be formalized. We must note that the logic of the content is still formal logic about conceptual form of the content, so there is no "logic of content" differing from formal logic in the deep nature. Of course, the natural language has been thousands of years, but the modern mathematical logic is only almost a hundred years, so it must first deal with the simpler problem. To illustrate the problem, we have a simpler example often used in the texts of the formalization of dialectic logic: "dialectical process of natural numbers $(0, 1, 2\ldots)$, negative integer, and integer". Now, we model the method in those texts: A as the "thesis" which is "S is a natural number", B as the "antithesis" which is "S is a negative integer", C as the "synthesis" which is "S is an integer", and domain is integer set. Of course, C is "S is either a natural number or a negative integer" in domain integer set.

3.6.2. *Chinese traditional approach*

The vast majority systems of formalized dialectical logic try to deal with dialectical negation by defining a non-classical operator and then use it to represent two "negations" at the same time in the negation of negation. As for us, the idea is too naïve, and we think the reason is that those Chinese and foreign logicians have not found the relationship between truth-functions and non-truth-functions. According to the basic principles of the STRF, there are only 16 basic (non-) truth-valued functional operators. But not any one of the 16 operators can meet these conditions at the same time: not only on the interest of the Hegels dialectics, but also on the uniform semantics of two "negations". So we conclude that, if we still maintain that dialectical logic can be formalized, two "negations" are only available with two different logical operators. In natural language or even in science language, the same word can express different concepts, which is the first day course in general logic teaching. The two "negations" here are the same word, but have not to represent the same logic operator.

We believe that having not noticed the impossibility of single logic operator can be used to represent two "negations" is the reason misguided most scholars who are in favor of formalizing dialectical logic.

3.6.3. *Dialetheias*

According to Graham Priest and Francesco Berto, *a* dialetheia *is a sentence (or a proposition),* A, *such that both it and its negation, $\neg A$, are true. Assuming the fairly uncontroversial view that falsity just is the truth of negation, it can equally be claimed that a dialetheia is a sentence which is both true and false. Dialetheism* *is the view that there are dialetheias. Therefore, dialetheism amounts to the claim that there are true contradictions. As such, dialetheism opposes the so-called Law of Non-Contradiction* (LNC) *that is, for any* A, *it is impossible for both* A and $\neg A$ *to be true, it can be regarded as a modern style of Hegels dialectical logic. Dialetheism has now become a live issue once more mainly because of (1) The logical (semantic and set-theoretic) paradoxes (2) Since the development of paraconsistent logic in the second half of the twentieth century.*

We only discuss a semantic paradox of self-reference in this sub-sub-section, and leave paraconsistent logic for next sub-sub-section. Though the semantic paradox is not the only example of dialetheias that have been mooted, it is exactly the most typical one. Prominent among them is the so-called Liar paradox. its standard version is following:

(1): (1) is false.

According to Priest and many other logicians *(1) refers to itself and tells us something about (1) itself. Suppose (1) is true: then what it says is the case, so it is false. Then, suppose (1) is false: this is what it claims to be, so it is true. If we accept the aforementioned Law of Bivalence, that is, the principle according to which all sentences are either true or false, both alternatives lead to a contradiction: (1) is both true and false, that is, a dialetheia, contrary to the LNC.*

At last sub-sub-section, we suggested that the same word can express different concepts in non-formal logic language, the situation here is similar that the same name appeared two times in natural language or science language, however, can still represent different propositions, this is the first day course in general logic lesson too. We order (1) first appeared as A, (1) second appeared as B, then "$A \equiv B$ is false", obviously, A and B are two different propositions. Suppose B is true: then "A = B is false = true is not true = false"; suppose B is false, then "A = B is false = false is false = true", obviously there are no contradictions. The cause that Graham Priest thinks there are contradiction is that there is a premise that he confused two different propositions A and B in which their truth-values are always negative each other into the same proposition. Of course, from a contradict premise, we can infer anything.

3.6.4. *Paraconsistent logic*

According to Graham Priest, Koji Tanaka, and Zach Weber, *a logic is said to be paraconsistent iff its logical consequence relation is not explosive. Paraconsistency is thus a property of a consequence relation,* $A, \neg A \models B$ *for every* A *and* B*(ex contradictione quodlibet (ECQ)).*

The most famous paraconsistent logic is Da Costa's "positive plus" systems. Fortunately, Priest had almost correctly noted the defect in Da Costa's systems (Priest and Routley 1989, pp. 164/5):

"*That an account of negation violates the law of non-contradiction provides prima facie evidence that account is wrong. This is [one] piece of evidence that da Costa negation is not negation.*

In fact we can make the claim more precise, traditionally A *and* B *are subcontraries if* $A \vee B$ *is a logical truth.* A *and* B *are contradictofies if* $A \vee B$ *is a logical truth and* $A \wedge B$ *is logically false. It is the second condition which distinguishes contradictories from subcontraries. Now in da 'Costa's approach we have that* $A \vee \neg A$ *is a logical truth. But* $A \wedge \neg A$ *is not logically false. Thus* A *and* $\neg A$ *are subcontraries, not contradictories. Consequently da Costa negation is not negation, since negation is a contradiction forming functor, not a subcontrary forming functor.*"

But unfortunately, Priest was wrong also with his dialethiest paraconsistency which B. H. Slater had correctly noted too[7].

"*Priest shows (Priest 1979, p. 238) that 'A is true' (*$V(A) \geq 0$*) is equivalent to 'A', and that 'A is false' (*$V(A) \leq 0$*) is equivalent to '*$\neg A$*'. But it is clear that 'A is true' and 'A is false' are not then contradictory, since what contradicts '*$V(A)0$*' is '*$V(A) < 0'$*. So 'A is true' is merely subcontrary to 'A is false', making '*$\neg$*' not a contradiction forming functor, in this logic.*"

In our opining, Priest inadvertently changed the definition of falsity which is much like the situation we mentioned in 3.4.3.

3.6.5. *Oppositional logic*

Probably effected by B. H. Slater's paper above and by Gui Qiquan, one of the leading logicians in paraconsistent logic and formalizing dialectical logic in China, who once was entangled with whether dialectical negation is implication or inverse implication in the early 1990's, Chinese logician Du Guoping invented an "Oppositional logic" that systemly researched intuitionistic logic, paraconsistent logic, classical logic, and even dialectical logic. He firstly published several articles about his "Metaphysical logic system" in Chinese in 2007, and then he wrote an abstract in English in which renamed his system as "Oppositional logic" in 2009[8].

Oppositional logic is an extended system of classical propositional logic. It can be obtained from the classical propositional logic by adding a unary connective $\star$ *and introducing the definitions of two unary connectives* $\triangle$ *and* ∇*. In oppositional logic system, there are four kinds of negation: the classical negation* $\neg$ *complying with both law of contradiction and law of excluded middle, the constructive negation* ∇ *complying with law of contradiction but not law of excluded middle, the paraconsistent negation* $\triangle$ *complying with law of excluded middle but not law of contradiction, as well as the dialectical negation* $\star$ *complying with neither law of contradiction nor law of excluded middle.*

[7]See Slater. B. H, 1995, Paraconsistent logic Journal of Philosophical Logic 24: 451–454.

[8]See Guo-ping DU, Logic, Rationality and Interaction, Lecture Notes in Computer Science, Volume 5834, 2009, p. 319.

Table 5. An Expansion of the Oppositional Logic.

A	f_1A	f_2A	f_3A	f_4A	f_5A	f_6A	f_7A	f_8A	f_9A
1	0	0	0	1	1	1	u	u	u
0	0	1	u	0	1	u	0	1	u

Note: $f_2(\mathbf{A}) = \neg\mathbf{A}$, $f_3(\mathbf{A}) = \nabla\mathbf{A}$, $f_6(\mathbf{A}) = \star\mathbf{A}$, $f_8(\mathbf{A}) = \triangle\mathbf{A}$, $f_7(\mathbf{A}) = \circ\mathbf{A} = \neg\triangle\mathbf{A} = \neg\star\neg\mathbf{A}$

In one of his three articles in 2007, he gave the proof of the soundness and completeness theorem of his system. Then it held the following conclusions in his English abstract in 2009:

[1] Oppositional logic can be a kind of tools for paraconsistent theory and intuitionistic theory; the famous Duns Scotus law does not hold according to the paraconsistent negation and the dialectical negation;

[2] In oppositonal logic, according to the unary connective $\neg$, $\star$, ∇ *and* $\triangle$, *A is in contradictory opposition with* $\neg A$*; A is in subaltern opposition with* $\star A$*; A is in contrary opposition with* ∇A*; A is in subcontrary opposition with* $\triangle$ *A. In this sense, we call the logical system mentioned above oppositional logic.*

$$\vdash A \to \star A; \triangle A \overset{def}{=\!=} \star\neg A, \quad \nabla A \overset{def}{=\!=} \neg\star A.$$

But in his most recent article in Chinese in 2011 as an expansion of his oppositional logic, he still thought that there are only 9 unary connectives in propositional logic with two-truth-valued as Table 5. ("u" means "non-definition")

Obviously, he did not realize that his "non-definition" in logical semantics is not as the genuinely "non-definition" discovered in our first research (Xiao-long Wan. 2011. The first exploration tounary operators: Whether can unitary operators be exhausted? Journal of Anhui University, social Science edn, in Chinese, 6th) but another sense of "definition" as our STRF showed. So, he did not know that "u" in Table 5 must have two different logical semantics in fact: 1/0 and 0/1, so that there are $4^2 = 16$ basic unary connectives in propositional logic with two-truth-valued as STRF showed, Furthermore he does not realize that his oppositional logic system is not an extended system of classical propositional logic but another sight of classical propositional logic itself according to STRF. That is, there is not any possible to not comply with either law of contradiction or law of excluded middle in his system.

3.6.6. *STRF approach*

According to STRF in Table 2, Du's $\star A$ is H_3A (or H_2A in other view) in fact, ∇A is $H_{14}A$ (or $H_{15}A$ in other view) in fact, $\triangle$ A is H_5A (or H_4A in other view) in fact, $\bigcirc A$ is $H_{12}A$ (or $H_{13}A$ in other view) in fact, and $\neg A$ is $H_{11}A$ in fact.

Let's consider the simple example in 3.6.1 about a two steps in negation of negation as a single dialectical negation process.

Step 1: from the thesis to the antithesis. In Hegel's sense, "from the thesis to its negation is further understanding of the thesis from the opposite side". In our model, "natural number" A is thesis, "negative integer number" B is the antithesis, and in the domain of integer number, "negative integer number" B is classical negation of "natural number" A. That is to say, Step1 is from A to B, $B = \neg A$,

Step 2: from the antithesis (i.e., the negation of the thesis) to synthesis. In Hegel's sense, "after the second negation, it seems if to be from antithesis back to the beginning of the thesis, but in fact, it is to achieve the synthesis which includes both the thesis and the antithesis". In our model, the antithesis is "integer number" C, "it seems if" is not "=" but "$\rightarrow$", so "it seems to be from antithesis back to the beginning of the thesis" can be interpreted as "$B \rightarrow C$". "It is to achieve the synthesis which includes both the thesis and the antithesis" can be interpreted as "$A \vee B$". That is to say, Step2 is from B to C, $C = A \vee B$.

For the domain is $A \vee B$, so $B = \neg A$, then the single dialectical negation process that is a negation of negation is from A to $\neg A$ to $A \vee \neg A$. Obviously, A and $\neg A$ is classical negation relation, $\neg A$ and $A \vee \neg A$ is subcontrary opposition relation, and A is in subaltern opposition with $A \vee \neg A$, just as Du's saying.

Moreover, in Hegel's sense, whole dialectical logic process includes many single negation of negation. We suppose that the domain is all number set, in the first negation of negation process: "natural numbers" A, "negative integer" B, and "integer" C, first negation is classic negation from A to $\neg A \neg A$ is "all numbers but not natural numbers"; second negation is paraconsistent negation which is H_5 in STRF as $\triangle$ in Du's, from $\neg A$ to C; B is included in the $\neg A$, $C = A \vee B$

Check: $A = 1/0$, $\neg A = H_{11}A = \neg(1/0) = 0/1$, $H_5\neg A = H_5H_{11}A = H_5(0/1) = 1/1//0$,

And $A \vee B = A \vee H_{14}A = 1/0 \vee 0/1//0 = 1/1//0$, so $C = H_5\neg A = H_3A$, $\vdash A \rightarrow H_3A$.

From above, we can see that one of the important characteristics of dialectical logic is that maintains the property of attribute of every related concepts (or propositions) unchanged, but by changing their domain, only changes the relation of the properties between the two sides of the dialectical negation, such that becomes the classical contradiction relation into non-classical contradiction relation, and leave room for to solve the contradictions. At the same time, STRF which deal with the non-truth valued function equivalent transforming to the truth valued function reflects the invariance of change of the dialectical logic from "informal", "inner", "reflexive", and "self-negation of content" to "formal", "external", "mutual", and "unary operator of negation".

And it is not very hard to formalize that Friedrich Engels' law of the unity of opposite, that is: the relation of A and H_5A or the relation of A and H_4A.

Probably Du Guoping is the first man who has written the basic syntactic form of Hegel's dialectical logic tightly, clearly, and almost correctly in the history, and STRF further advanced it in a more general way and give it the semantics precisely and exactly.

Hegel is dear a great man. Before modern mathematical logic invented one or two centuries ago, he discovered the dialectical logic. Even though he could not understand his logic distinctly, whose every dialectical negation process from "thesis" to "synthesis" is exactly the implication of classical logic, but the two steps of inner dialectical negation process from "thesis" to "antithesis" and from "antithesis" to "synthesis" are not always abide by the implication rule of classical logic, which reflects the internal dynamic and organic fine structure of human logical reason.

4. Conclusion

The principle of equivalent transformation between non-truth-function and truth-function is simple but profound. Especially, according to STRF no strictly defined, we can automatically generate the Failure of Equivalent Substitution Principle. According to it formal defined, we have to obtain Incompletely Substitutability Rule and Distribution Feather of Truth-valued Degree totally in classical logical way to solve many problems once in non-classical logic way. No matter how complex and mysterious is the counterpart in natural language and in scientific language corresponded by some kind of non-truth-function operator, which formed the so-called "non-classical logic" that has to be an "idiom" in classical logic in our opinion. Only one operator and their superposition can make any non-truth-function equivalent to a truth function in STFR, which is not the case that countless unary operators make all non-truth-functions equivalent to countless every truth functions. Only can six symbols form any proposition, which does not only represent formal unification, but also conform to the characteristics of natural language. On the other hand, single non-truth-functional connective in natural language or in scientific language, "obligatory", "ought to", "dialectical negation", or "$\sqrt{not}$", etc, turns out to be a supposition of several classical or non-classical unary operators which can be equivalent to a classical truth-valued function with many-variables, which shows that the natural language is actually "save" and "safety". And the same two non-truth-valued function operators shown in a supposition (e.g. HHp) is just the origin of holism, only one variable with other five symbols can form any proposition is the firm and formal base of a new and precise monism. STRF focus only some "non-classical" problems which come from the corresponding relations between non-truth-functional operates and binary truth-valued function, other "non-classical" problem will be researched by more general theory in future.

Though completely convinced explanting STRF needs a theory of the truth in well developed, the establishment of STRF is not dependent on any philosophical theory of truth, which is only based on the unification between the pure formal universality of symbol logic and the adaptability of the natural language. By contrast, there are many logicians have been blinded by vagueness and ambiguity on the surface of the natural language and do not realize its "save and safety" in the empirical appropriateness. And many philosophers who are over-dependent on some reference theory or special truth theory in analytic philosophy forget the independence of universality in the formal nature of logic. The formal nature of logic is dichotomy, so that modern classical logic is the simplification of mathematics and clarification of philosophy extremely.

References

Anderson, A. R. 1967. Some Nasty Problems in the Formal Logic of Ethics. Noûs 1(4): 345–360. 1962. Conditional Permission in Deontic Logic. Philosophical Studies 13(1–2): 1–6. 1958. A Reduction of Deontic Logic to Alethic Modal Logic. Mind 67(265): 100–103.

Barcan, R. C. 1946. A Functional Calculus of First Order Based on Strict Implication, Journal of Symbolic Logic 11: 1–16.

Belnap, Nuel D. *et al.* 2001. Facing the Future: Agents and Choices in Our Indeterminist World. Oxford University Press on Demand.

J. van Benthem and Sonja Smets. 2011. New Logic Perspectives on Physics, Synthes (DOI 10):1007/s11229-011-9911-y, published online.

G. Birkhoff and J. von Neumann, 1936. The Logic of Quantum Mechanics. Annals of Mathematics 37: 823–824.

Blackburn, P., de Rijke, M. and Venema, Yde. 2002. Modal Logic, Cambridge University Press.

Blackburn, P., van Benthem, J. and Wolter, F. 2007. Handbook of Modal Logic, Elsevier.

Burgess, J. P. 1999. Which Modal Logic is the Right One? Notre Dame Journal of Formal Logic 40(1): 81–93.

Carnielli, W. and Pizzi, C. 2008. Modalities and Multimodalities, Logic, Epistemology and the Unity of Science 12, Springer.

Chen Bo. 2008. Thinking Deeply, Contributing Originally: An Interview with Timothy Williamson. Studies in Logic 1(3): 79–109.

Cheng-Chih Tsai, 2012. The Genesis of Hi-Worlds: Towards a Principle-Based Possible World Semantics. Erkennis 76: 101–114.

Chiara, M. L. D. *et al.* 2004. An Unsharp Logic from Quantum Computation, International Journal of Theoretical Physics 43: 1803–1817.

Cocchiarella, N. B. and Freund, M. A, 2008. Modal Logic: An Introduction to Its Syntax and Semantics, Oxford University Press.

Copeland, B. J. 2002. The Genesis of Possible Worlds Semantics, Journal of Philosophical Logic 31: 99–137.

Copeland, B. J. 2006. Meredith, Prior, and the History of Possible Worlds Semantics, Synthese 150: 373–397.

Niels Egmont Christensen, 1965. A Non-Truth-Functional Interpretation of Mathematical Logic, Analysis 25: 129–132.

Deutsch, D. 1992. Quantum Computation. Physics World, June, 57–61.

Divers, J. 2006. Possible-Worlds Semantics without Possible-Worlds: The agnostic Approach. Mind 115: 187–226.

Kurt Engesser *et al.* eds, 2009. Handbook of Quantum logic and Quantum Structures, Elsevier.

Gabbay, D. M. & F. Guenthner (eds.) 2002. Handbook of Philosophical Logic, 2nd Edition. Kluwer.

Garson, J. 2006. Modal Logic for Philosophers, Cambridge University Press.

Garson, J. 2005. Unifying Quantified Modal Logic,? Journal of Philosophical Logic 34: 621–649.

Goble, L. 2000a. The Concept of Moral Obligation. Philosophy and Phenomenological Research 60(1): 242–244. 2000b. Multiplex Semantics for Deontic Logic. Nordic Journal of Philosophical Logic 5(2): 113–134. 1993. The Logic of Obligation, 'Better' and 'Worse'. Philosophical Studies 70(2): 133–163.

Goldblatt, R. 2006. Mathematical Modal Logic: A View of Its Evolution, in Dov M. Gabbay and John Woods eds, Handbook of the History of Logic, Volume 7, pp. 1–98.

Guo-ping DU, Logic, Rationality and Interaction, Lecture Notes in Computer Science, Volume 5834, 2009, p. 319.

Haack, S. 1978. Philosophy of Logics, Cambridge University Press.

Hill, Daniel J. & Stephen K. McLeod, 2010. On Truth-Functionality, Review of Symbolic Logic 3(4): 628–632.

Horacio Arlo-Costa Eric Pacuit, 2006. First-Order Classical Modal Logic, Studia Logica 84: 171–210.

Hughes, G. E. and Gresswell, M. J. 1996. A New Introduction to Modal Logic, London and New York: Routledge.

Jackson, F & J. E. J. Altham. 1988. Understanding the Logic of Obligation. Aristotelian Society Supplementary Volume 62: 255–283.

Jackson, F & Robert Pargetter. 1986. Oughts, Options, and Actualism. Philosophical Review 95(2): 233–255.

Jackson, Frank. 1985. On the Semantics and Logic of Obligation. Mind 94(374): 177–195.

Kolak, D. 2001. On Hintikka, Wadsworth.

Kolak, D. and Symons, J. eds., 2004. Quantifiers, Questions and Quantum Physics: Essays on the Philosophy of Jaakko Hintikka, Springer.

Korte, T., Maunu, A and Aho, T. Modal logic from Kant to Possible Worlds Semantics, in The Development of Modern Logic, ed., Leila.

Haaparanta, Oxford University Press 2009.

Kripke, S. A. 1965. Semantical Analysis of Modal Logic: Non-Normal Modal Propositional Calculi, Symposium on the Theory of Models; North-Holland Publi. Co., Amsterdam. 1963, Semantical Considerations on Modal Logic, Acta Philosophica Fennica 16: 83–94. 1959. A Completeness Theorem in Modal Logic, The Journal of Symbolic Logic 24: 1–14.

Lewis, C. I., 1912. Implication and the Algebra of Logic, Mind 84: 522–531. 1914. The Calculus of Strict Implication, Mind 90: 240–247. 1920. Strict Implication–An Emendation, The Journal of Philosophy, Psychology and Scientific Methods 17: 300–302.

Li Xiao-Wu, Modal LogicZhongshan Univ. 2006.

Lukasiewicz, 1930. Philosophical remarks on many-valued systems of propositional logic [J], in Jan Lukasiewicz selected works, edited by L.

Borkowski, North-Holland Publishing Company, 1970, p. 153.

Lukasiewicz, On Three-Valued Logic [J], in Jan Lukasiewicz Selected Works, edited by L. Borkowski, North-Holland Publishing Company-Amsterdam London, 1970, p. 88.

Marcos, Joäo, 2009. What is a non-truth-functional logic?, Studia Logica 92(2): 215–240.

James A. Marten. 1973. Are There Truth Functional Connectives? Metaphilosophy 4: 187–204. McCarthy, J. 1997. Modality Si! Modal Logic, No! Studia Logica 59: 29–32.

McKeon, M. 2005. A Defense of The Kripkean Account of Logical Truth in First-Order Modal Logic, Journal of Philosophical Logic 34: 305–326.

McNamara, P. 1996. Doing Well Enough: Toward a Logic for Common-Sense Morality. Studia Logica 57(1): 167–192. 1997. Supererogation in Deontic Logic: Metatheory for DWE and Some Close Neighbours. Studia Logica 59(3): 397 2004. Review: Agency and Deontic Logic.? [REVIEW] Mind 113(449): 179–185.

Ming Xu, An Intruduction to Symble Logic (in Chinese), Wuhan Univ. 2008.

Moretti, A. 2009. The Geometry of Standard Deontic Logic. Logica Universalis 3 (1).

Paul McNamara, 2006. Deontic Logic. Stanford Encyclopedia of Philosophy.

Peregrin, J. 1993. Possible worlds: A Critical Analysis, in Prague Bulletin of Mathematical Linguistics 59–60: pp. 9–21.

Pollard, S. 2005. The Expressive Unary Truth Functions of n-valued Logic, Notre Dame Journal of Formal Logic 46(1): 93–105.

Francesco Berto & Graham Priest. 2008. Dialetheism, The Stanford Encyclopedia of Philosophy.

Graham Priest, Koji Tanaka and Zach Weber, 2013. Paraconsistent Logic (Stanford Encyclopedia of Philosophy)

G. Priest, R. Routley & J. Norman (eds.) 1989. Paraconsistent Logic: Essays on the Inconsistent, Philosophia Verlag

Quine, W. V. 1947. The Problem of Interpreting Modal Logic, The Journal of Symbolic Logic 12: 43–48. 1961. Reply to Professor Marcus. Synthese 13(4): 323–330. 1976. The Ways of Paradox, and Other Essays. Harvard University Press. 1982. Methods of Logic. Harvard University Press

Rescher, N. 1999. How Many Possible Worlds Are There? Philosophy and Phenomenological Research 59: 403–420. 1961, The Concept of Randomness, Theoria 27(1): 1–11.

Schnieder, B. 2008. Truth-functionality, Review of Symbolic Logic 1(1): 64–72.

David Sherry. 1999. Note on the Scope of Truth-Functional Logic. Journal of Philosophical Logic 28: 327–328.

D. J. Shoesmith and T. J. Smiley, 1971. Deducibility and Many-Valuedness, The Journal of Symbolic Logic 36(4): 610–622 p. 611.

Slater. B. H, 1995. Paraconsistent logic? Journal of Philosophical Logic 24: 451–454.

Stephen Read. 1992. Conditionals are not Truth-Functional: An Argument from Peirce. Analysis 52: 5–12.

von Wright, G. H. 1999. Deontic Logic: A Personal View. Ratio Juris 12(1): 26–38. 1981a. Deontic Logic and the Theory of Conditions. In Risto Hilpinen (ed.), Deontic Logic: Introductory and Systematic Readings. Sold and Distributed in the U.S.A. And Canada by Kluwer Boston. 1981b. A New System of Deontic Logic. In Risto Hilpinen (ed.), Deontic Logic: Introductory and Systematic Readings. Sold and Distributed in the U.S.A. And Canada by Kluwer Boston. 1968. An Essay in Deontic Logic and the General Theory of Action. Amsterdam, North-Holland Pub. Co.. 1967. Deontic Logics. American Philosophical Quarterly 4(2): 136–143. 1956. A Note on Deontic Logic and Derived Obligation. Mind 65(260): 507–509. 1951. Deontic Logic. Mind 60(237): 1–15.

Vranas, Peter B. M. 2010. In Defense of Imperative Inference. Journal of Philosophical Logic 39 (1). 2008. New Foundations for Imperative Logic I: Logical Connectives, Consistency, and Quantifiers. No?s 42(4): 529–572.

Williams, B, 1995. Truth in Ethics. Ratio 8(3): 227–236.

Xiao-long Wan. 2011. The first exploration to unary operators: Whether can unitary operators be exhausted? Journal of Anhui University (social Science edn, in Chinese), 6th.

Xiao-long Wan *et al.* 2012. The second exploration to unary operators: Deontic logic and deontic paradox, Journal of Anhui University (social Science edn, in Chinese), 3th.

Xiao-long Wan. 2012. The third exploration to unary operators: On modern modal logic from STRF. Journal of Huazhong University of Science and Tech. (social Science edn, in Chinese), 3th.

Xiao-long Wan *et al.* "On G. Birkhoff and J. von Neumann's Quantum Logic- The forth exploration to unitary operators", in Journal of The Studies in Philosophy of Science and Technology (in Chinese), 2013.

MECHANISM AND PRODUCTIVE EVENT

ZHU XU

Department of History and Philosophy of Science
University of Chinese Academy of Sciences
Beijing 100049, China
E-mail: dennis.zhxu@ucas.ac.cn
people.gucas.ac.cn/~xuzhu?language=en

Production is an essential notion for theory of mechanism, and to clarify its meaning brings about lots of difficulties. This article argues for an event-property theory which characterizes production as a particular property of event. That theory may provide a prospective solution for those difficulties.

Keywords: Mechanism; Causation; Production; Event-property theory.

1. Introduction

Humean regularist theory has been a standard view for causation in most years of 20th century. In the last two decades, philosophers of science put more and more attentions on causation and explanation in special sciences. Some of them claim interventionist theory which conceptualizes causation in terms of manipulability among variables, as "invariance under intervention"(Hausman and Woodward 1999; Woodward 2000). Others are mechanistic theorists, who argue that the notion of mechanism, though differs from mechanistic philosophy in 17th century, implies causation in special sciences, such as biology, neuroscience, and the social sciences (Glennan 1996, 2002; Machamer, Darden, and Craver 2000). Though both are critics on the regularist view, interventionist and mechanist theories are still competing with each other (Machamer 2004; Psillos 2004; Woodward 2002, 2011).

The essential problem here is how to characterize production. The mechanist theory appeals to a general intuition that causes play some productive roles on effects. Mechanist theory claims production could not be correctly represented from the notion of causal dependence; while the interventionists insist that there is no distinctiveness of production other than dependence among variables. Mechanist theory could only provide detailed spatial-temporal information about "invariance under intervention", rather than any distinctive position on causation. Therefore, some characterization which could elucidate the distinctiveness of production would be necessary for the justification of mechanist theory.

This article opens with a brief introduction of the mechanist theory, which demonstrates its competitions with interventionist theory. From the second section, it turns to the characterization of production. I will argue that those attempts till now are trapped in a serious dilemma. My suggestion is to resolve the dilemma with a particular theory of production, the event-property theory. I would argue for how the theory characterizes production, and in further, why it makes contributions to the resolution of those difficulties faced by other attempts.

2. Mechanism and Production

Many philosophers have recognized mechanism as a popular notion especially employed in special sciences. However, what the notion means is still controversial. How can we characterize the essence of mechanism? One of the points of view is to claim that causal laws are involved in the very essence of mechanism:

"A mechanism underlying a behavior is a complex system which produces that behavior by of the interaction of a number of parts according to direct causal laws."
(Glennan 1996, p. 52)

According to Glennan's definition, mechanism, causation, and laws of nature seem probably to be unified into a comprehensive understanding. Glennan believes it is worthwhile to argue for mechanism taking a significant status in special sciences. For instance, when they talk about inquiries of causation, molecular biologists usually mean to identify the substantial mechanism which brings about the phenomena.

Another view is to assume that the identification of mechanism doesn't need to involve laws of nature. Peter Machamer, Lindley Darden, and Carl Craver (hereafter MDC) put forward a different characterization of mechanism.

"Mechanisms are entities and activities organized such that they are productive of regular changes from start or set-up to finish or termination conditions."
(Machamer, Darden, and Craver 2000, p. 3)

Mechanism is identified as organization of entities and activities. Entities are supposed to carry out those substantial changes in mechanism, while activities are producers of those substantial changes in which entities participate. And mechanism is a particular spatial-temporal organization in which entities are involved in certain activities. For example, "entities in neurobiology include such things as pyramidal cells, neurotransmitters, brain regions, and mice. ...Pyramidal cells *fire*, neurotransmitters *bind*, brain regions *process*, and mice *swim* in water while eagerly *searching* for a means of escape ... Activities are the things that entities do; they are the productive components of mechanisms, and they constitute the stages of mechanisms."(Craver and Darden 2001) Special verbs could attract much attention on possibilities of representing activities. That is an extremely important feature of mechanism which will receive more discussion in the next section.

Mechanism is supposed to be "productive of regular changes". Mechanism is not a single process that only occurs in a certain region and particular timing, but rather a phenomenon which could in principle occurs regularly under the same conditions. By the same reason, scientists actually talk about mechanical models when they mention "mechanism".

The productive relation is usually described as continuity of the temporal organization of mechanism. A complete description of mechanism should illustrate itself as a coherent causal narrative without any obscure interruption. MDC gives an example on productive continuity by the mechanism of protein synthesis. How could DNA control protein synthesizing? Now scientists explain it with "mechanism schema" involving activities such as transcription and translation. The most significant discovery is of the role played by RNA in the mechanism. It turns out that RNA takes part in transcription of DNA in one aspect, and then translates the heritable information to protein synthesis in another aspect. Therefore, DNA-RNA-protein synthesis has been constitutive to a coherent causal narrative without any explanatory gaps. When the role of RNA was still unclear, scientists cannot be satisfied and have to inquire for the intermediate link from DNA to proteins. Undoubtedly, the schema above is very abstract and there are still many details in the mechanism should be fulfilled to provide a complete causal story, and then the productive continuity will become more informative.

Temporal organization is a necessary condition for mechanism. Productive continuity suggests that mechanism is explicitly temporal-organized stage by stage. The duplication of DNA, the transcription of RNA, and translation of heritable information should be arranged in a certain order. Spatial organization is also necessary for mechanism. Mechanism always occurs in a specified region. And it is usually influenced by those particular spatial features and structures. For instance, neurobiologists now have identified the Long-Term Potentiation (LTP), located in the synapses of the hippocampus of brain, as the main mechanism for spatial memory. (cf. Craver and Darden 2001, especially pp. 115-119) Because of the particular spatial structure, namely that the pre-synapse cell is confined to the boundaries of cell membranes and then separated from the post-synapse cell, the relevant study for LTP should pay much attention on the possible connection in the synapse, which provides guidance for explaining the LTP mechanism by focusing on neurotransmitters. Scientists have admitted in their causal inquiry to specify mechanism by some strategies from localization and decomposition. (Bechtel and Richardson 2010)

One of critical points in MDC's conceptualization is to claim that laws of nature are not necessary for mechanism. It is by this point of view that MDC's notion of mechanism is distinguished from Glennan. There is no need to assume laws of nature underlying productive activities, since it is the production rather than universality that defines causation in special sciences. Universality or high probability may be the most important concern for causal inquiries in physical science, while in special sciences scientists may focus on actual production, which may not meet with requirements from universality, and may involve numerous implicit *ceteris paribus* clauses. For example, the transcription of RNA may introduce no universal law, but it is the actual occurrence by which the link from DNA to protein synthesis is accomplished. And that is the very sense of causality in special sciences, which would not be illustrated by universal laws.

However, MDC's conception of mechanism has been faced with serious objections. Some critics believe that it is untenable to identify mechanism without any laws. (Leuridan 2010) Others who are proponents for the interventionist theory argue that the notion of mechanism should take counterfactual dependence as a necessary condition. Psillos (2004; 2011) claims on the interventionist view as a more fundamental intuition of causality than the notion of mechanism. Woodward (2002, p. S370) puts forward some particular argument rightly on this point:

"If a generalization *G* relating *X* to *Y* is to describe a causal relationship (or in MDC's language), a 'productive' relationship, it must be invariant under at least some (but not necessarily all) interventions on *X* in the sense that *G* should continue to hold (or to hold approximately) under such interventions."

Woodward seems to believe that the productive activities, interpreted as intrinsic character of mechanism by MDC, are merely another elaboration of "invariance under interventions". Mechanist theory is thought of as merely to provide some spatial-temporal information for the manipulability relationship (Woodward 2011, p. 424). Interventionists insist that, variables correlation plays an indispensable role in causal inquiries of special sciences,[a] which has replaced any independent status of mechanism.

Some philosophers intend to defend an independent intuition of production. Whether mechanist theory could be reasonably distinguished from interventionist view depends on how to characterize production. It is production that maintains the distinctiveness of mechanism from variables correlation. MDC has already given a tentative characterization on production.

3. Production as Verb-formed Activity

Production usually implies active powers assumed in the productive process. For instance, Tom kicks a ball and the ball is rolling away. Tom "produces" the ball to roll by his kicking. Kicking is a particular active power actually operated in that production. And it has ruled out those situations which Tom causes the ball to roll away in any other way, for example, Tom pushes a gearing device, which in further motivates the ball to roll ahead.

The distinctiveness of production could be represented in special verbs relative to particular entities. For example, "kicking" is a proper verb-form for the production involving entities like persons and balls. Verb-forms point out a possible approach to characterize production. Many verbs are employed in the platitudes of causation, such as pushing, pulling, kicking, scratching, striking, and driving. It seems that the word "cause" is too general to be informative about what kind of causation in effect. Cause is not a single notion but a set of notions which illustrate various particular kinds of causation in verb terms (Anscombe 2008; Cartwright 2007).

MDC (2000, p. 4) puts forward a characterization of production in terms of verb-formed activities. "An activity is usually designated by a verb or verb form (participles, gerundives, etc.). Activities are the producers of change. They are

[a] We could elucidate some similarities between regularist and interventionist theories on the notion of dependence among variables. Regularity may be interpreted as dependence of a unique sort, which is defined as the same phenomenon occurs repeatedly. In this sense the interventionist theory claims only another specified dependence, sort of counterfactual one among variables. That observation seems to make sense of the interventionist objection to mechanism as a refined edition from the regularist one: Although universal laws are not necessary, the dependence among variables should be an unavoidable presupposition for causation.

constitutive of the transformations that yield new states of affairs or new products."

Just like kicking is a right activity for entities like persons and balls, some verbs in senses specified by sciences could also be specified as essential activities of particular entities assumed and referred in scientific theory. Here is a case given by MDC on characteristic activities for theoretical entities. "Attracting, repelling, bonding, and breaking are electro-chemical kinds of activity. ... The historical development of the mechanism of protein synthesis required finding an activity to order linearly the constituents of the protein, its amino acids; an early idea using primarily geometrico-mechanical activities was replaced by one involving, primarily, the weak electro-chemical activities of hydrogen bonding." (Ibid., p. 14)

Especially, Machamer claims an improved edition of the verb-form theory of production. First, production is primarily thought of as actuality rather than potentiality. Production in its proper verb-form should be actual occurrence. When we talk about the production elucidated in Tom's kicking of the ball, what we mean is not just that the kicking involves Tom's active power, but also that the power as potentiality has already been actualized by his kicking. Though active power seems to be employed in any productive process, production is still essentially presumed by active power, as actuality is logical prior to potentiality (Machamer 2004, p. 30). It seems clear on the distinction between the verb-formed conception and those accounts of causation involving notions such as capacity, disposition, tendency, and so on, which could be roughly unified into the notion of causal power. Many causal realists maintain theories of causation constitutive from those notions of potentiality (Cartwright 1989; Harré and Madden 1975; Shoemaker 2003). However, the verb-form theory commits to a different ontology from those causal realists. Production is categorized as local and real occurrence in the uniquely actual world, rather than any capacity or tendency of which meanings may involve considerations about possible worlds.

Furthermore, since it is considered as occurrence of actuality, production does not involve any sort of ontological necessity. In fact, Machamer (2004, p. 31) even seriously doubts whether there is any necessity of an ontological kind. Machamer argues, when we apply an intuition about causality, it could be identified in the productive sense without any assumptions about ontological necessity. If the identity of cause is a producer of the event, causation could be considered as the productive occurrence in reality, without any appeal to possible worlds. While Hume attempts to deny ontological foundation in order to disprove causality, Machamer believes that there is no need in principle to assume ontological necessity for causality. As Machamer points out, we may only appeal to epistemic or methodological notion of necessity, rather than an ontological one.

However, there is no argument for that the ontological picture should be different from the epistemic or methodological one. What if they are the same? For instance, the core of causation, according to the interventionist theory, is "difference-making" information, which identifies cause as a difference-maker on the value of effects. Why is this only an epistemic or methodological elucidation rather than an ontological one? "Mechanistic explanation involves, so to speak, difference-making information all the way down, but spatio-temporally organized difference-making information." (Woodward

2011, p. 424) If process of production is just spatio-temporal organization of difference-making, then there is no substantial independence for mechanist theory from interventionism.

Finally, production as verb-formed activity cannot reduce to canonical or functional representations[b]. Interventionist theory depends crucially on canonical and functional representation of causation. The chemical bonding, e.g. that amino acids are formed to strong covalent bonds, produces the protein synthesis. A canonical representation would first give at least two variables on the bonding of amino acids and the protein synthesis respectively, and then describe the correlative dependence (maybe counterfactually) among variables. However, in Machamer's view, this cannot provide any ontological picture of the real and local causation from amino acids to the synthesized protein, since it says nothing about the intermediate production. The causal dependence could not warrant that amino acids are bonded to be protein synthesis. Consider the case of Tom's kicking: we can determine the dependence between Tom stretching his leg ahead and the rolling of ball, while Tom does not kick the ball. He makes the ball to roll by a gearing device. Things are similar in the scientific case: Chemical bonding is a particular activity. Even if the causal dependence could be identified epistemically, there is no qualification on whether this dependence is realized by the real and local bonding or not. Production could not be treated as merely a spatial-temporal organization of difference-making information. It is the verb-formed activity that does the difference-making job, while its particular qualification of production could be represented neither by causal dependence among variables, nor the spatial-temporal organization of that dependence.

The first problem for the verb-form theory is that there are lots of similar verbs which merely describe activities superficially. Not all verbs are appropriate for illustrating production relatively to contexts. For example, we could say "Tom activates the ball to roll" instead for Tom kicking the ball. However, few people would admit that Tom's "activating" of the ball has described an activity, since it doesn't give out enough information for production[c]. Mechanist theory could not be satisfied on existence of production, but is supposed to identify the production of particular kind, which is defined in qualified proper verbs. However, Machamer has not made a distinction between those appropriate productive verb-form and other verbs superficially for characterizing production.

Second, it seems very peculiar to hold merely epistemic and methodological necessity rather than ontological one. From views of platitudes and interventionism, modality is an indispensable implication of causality. However, verb–form theory only

[b] "Natural languages need verbs (or verb surrogates), and some verbs refer to causes or activities. In formal languages, verbs may be eliminable in some sense (in favor of functions). Canonical logical form eliminates verbs and treats them as strange predicates or properties of objects. This seems to be equivalent to turning all verbs into property-like capacities, and I have argued above against the sufficiency of wantonly nominalizing everything. Moreover, taking functions as ontologically basic would yield an even weirder ontology than countenancing activities. Logic then would have become metaphysics." (Machamer 2004, p. 31)

[c]Craver (2007, p. 113) put those superficially productive verbs as "filler terms that give the illusion that the explanation is complete when it is not". Those verbs "are often used to indicate a kind of activity in a mechanism without providing any detail about exactly what activity fills that role".

respects actuality as intrinsic feature of production, which may bring about doubts on whether it could be an acceptable characterization of causation thoroughly sunk in modality.

The final worry for verb-form theory comes from maintaining the irreducibility to canonical representations. Canonical representation is unavoidable in scientific practice. It is characteristic for scientific representation that every process should be described by dependent correlations among variables, which is a requirement from Galileo's mathematical idealization of natural world. If the irreducibility means impossibility for dependence among variables to represent production, we cannot get any valuable scientific representations. And that's obviously not true in scientific practices.

4. The Dilemma of Characterizing Production

The last objection may implicitly point to a reductive analysis of production. Hall (2004, p. 265) provides us very clear possibility for that reduction, which aims to save the notion of production out of the status "as an unanalyzable causal primitive". Hall's reductive analysis of production makes no commitment to interventionism. Interventionist theory does not admit any independent status for production. Nevertheless, Hall claims production and dependence are different notions of causation[d]. Production is irreducible to dependence among variables, but is indeed reducible to canonical representations.

In Hall (2004, p. 225), the notion of production is defined by three main features:

"*Transitivity*: If event c is a cause of d, and d is a cause of e, then c is a cause of e.

Locality: Causes are connected to their effects via spatiotemporally continuous sequences of causal intermediates.

Intrinsicness: The causal structure of a process is determined by its intrinsic, non-causal character (together with the laws)."

On the contrary, the notion of dependence involves counterfactual dependence and the feature so-called *Omission*, which specifies "failures of events to occur" as relevant terms in a causation. Both notions are independent to each other, since Hall argues that we can have a causal case in principle well-designed for satisfying *Transitivity*, *Locality*, and *Intrinsicness*, while it cannot be intelligible from counterfactual dependence or *Omission*; and that there is another case which could only satisfy the notion of dependence rather than production[e].

For Hall, the last objection to verb-form theory is really serious: to deny reducibility seems to eliminate analyzability of verb-formed activity. Thus the irreducibility position

[d]Hall's claim about two notions of causation rises from considerations on the puzzlement of counterfactual analysis of event causation (Bennett 1993; Lewis 1986; Lombard 1990).

[e]Neither of them could be sufficient for causation. But they are complementary. For example, Hall points out that the over-determination case, which is thought of as a serious problem for counterfactual analysis, is well-satisfied by the notion of production. And he also put forward the double prevention case, which is qualified as causal by the notion of dependence, to show that production is not yet sufficient for causation. (Hall 2004, p. 253)

should be refused at all. That has been expressed in the *Intrinsicness* thesis. If the intrinsic character of production, which determinates its causality, is non-causal *per se*, production then could not be primitive notions of the world, since it is in principle analyzable in some non-causal representative ways. And that is exactly what Hall intends to argue. Verb-formed activity is reducible to canonical representation; and the causal is supervenient on the non-causal.

Therefore, though he commits to the distinction between production and dependence, Hall does not agree totally with Machamer. His analysis is definitely not to replace production with counterfactual dependence among variables; however, production in his analysis is reducible to canonical representations. It seems that both Machamer and Woodward are implicitly convinced on some conceptual relation between canonical representation and counterfactual dependence. And that is absolutely what Hall intends to deny. Production, in non-causal canonical representation, could still be different from counterfactual dependence. For example, those theses of *Transitivity*, *Locality*, and *Intrinsicness* are derived from canonical representations of production, and they cannot be satisfied by counterfactual dependence. It would rather say that production is parallel to dependence, and both of them could be put under canonical representation.

However, there is still a serious challenge to the reductive analysis of production. The reason for irreducibility is that canonical description does not provide enough details – especially the kind of activities – about the production. As required by scientific representation, we should first stipulate some variables to represent relevant causal terms, and then investigate the correlation among those variables. Machamer argues that, even dependence among variables has been specified, we may still have no idea on the particular kind of the intermediate productive process. The reductivist view may response to this challenge by insisting the distinction between production and dependence under canonical representation. However, it cannot resolve the challenge, since what the irreducibility challenges is actually the intrinsic character of canonical representation. Can we find detailed information about particular kind of production in the canonical representations? Even if it is specified that, assuming some appropriate variables, the synthesis of proteins satisfies all canonical features of production, such as *Locality*, *Transitivity*, and *Intrinsicness*, we still have no idea about whether chemical bonding is the right kind of activity for amino acids to synthesize proteins. In order to elucidate the production, it is significant to distinguish geometrico-mechanical from electro-chemical activities, which seems impossible under the canonical representations.

In summing up the attempts from Machamer and Hall, we could find a serious dilemma on characterizing production. One horn is that, if we hold the irreducibility to canonical representation, production cannot get rid of black-box status, or uninformative "filler terms", by Craver's notion. This is where Machamer's point falls. Nevertheless, if we make explicit production by canonical representations, it is actually put parallel to counterfactual dependence, which is in risk to lose the distinctiveness of production. That is the second horn in which Hall's reductive analysis is trapped.

5. Production as Property of Event

It is likely that the only hope of resolving the dilemma depends upon how to characterize production. Here we need a study for productive metaphysics. The metaphysical study of production should avoid those two dispositions. The first is to the ultra-metaphysics of production, which takes human agency – or anything constructed by agency as constitutive element of production. Production, in the ultra-metaphysical picture, is illustrated as something very different from other parts of nature in the scientific view. On the contrary, the second one is minimal-metaphysics of production, according to which production refers only to the empirical phenomena best explained by it, while loses nearly all the independent meaning from canonical representations.

Therefore, what we need is moderate metaphysics of production. The first problem of moderate metaphysics is whether production is substance or property. It seems untenable to believe production as particular kinds of substance. If so, production may probably be particular properties. For example,

(*PP*) "RNA translates the information from DNA into protein synthesis."

Could production be a property of facts or states of affairs? *PP* describes a state of affairs, say, *Trans (RNA, DNA, protein)*. The relational fact is a truth-maker for *PP*. But production could not be a property of the truth-maker fact, since states and relations are not appropriate substances for productive properties. Production should be continuity in spatial-temporal organization. However, states of affairs or relations like *Trans (RNA, DNA, protein)* could only hold a particular point in the 4-dementional space-time, while substances of production should involve at least continuity among different points.

That leads us to think about the possibility that production is a property of event. Actually, *PP* means there is a productive event of RNA translation. Any particular event involves a spatial-temporal continuity, which seems to be more suitable for substances of production than facts or relations.

Event theory is seriously controversial among philosophers (Davidson 1980; Horgan 1978; Kim 1976; Bennett 1996). Here the only thing I care about is substantial change in event, which is intimately related to production. Some philosophers associate event with the notion of change (Lombard 1979; Cleland 1991), which provide direct foundation for event-property theory, the characterization of production I want to argue.

(*PD*) An event *E* is productive, only if:

i. *E* is a causal event located on a particular temporal-spatial region;

ii. *E* involves at least one concrete change *R* consist of original and terminal states. And *R*'s terminal state is a somewhat novelty in terms of contextual elements;

iii. *R* always refers to verb-forms or material arrangements for intervention.

Productive property would not give any non-circular definition for causality. We cannot determine an event as productive or not without specifying it as causal. Usually we describe an event as causal when it is specified to involve productive properties. It is not surprising that the inquiry for productive property is the same thing as the inquiry for causal relation, especially in, for example, biology and cognitive neuroscience. Conceptual relations between production and causation are intrinsically constructive for each other, which deny any reductive definition[f].

The first thesis indicates that productive property occupies temporal-spatial continuity. That occupancy is one of the distinctiveness of event from state of affairs. It should be noticed the way that event occupies temporal-spatial region is different from physical object. For instance, I'm currently thinking about writing an essay. Thus at the same time there occurs a neurological event and a psychological event in my brain. One single region is not supposed to accommodate two different objects, while it could contain different events.

Productive property particularly specifies "concrete change" of event. The notion of concrete change presumes a distinction between "phase" and "state". "A state is a determinate property (or *n*-adic relation) … A phase is a determinable property (or *n*-adic relation)." (Cleland 1991, p. 233) A single phase can correspond to various states. For instance, a car is in a velocity of 80 miles per hour. The velocity of car is a phase, which is actually but not necessarily in accordance to a particular state, 80 miles per hour. "A concrete change R is a pair $\{x, y\}$ such that x is the exemplification of a state s by a concrete phase CP at a time t and y is the exemplification of a state s' by a concrete phase CP' at a time t', where (i) t is earlier than t'; (ii) CP is the same concrete phase as CP'; and (iii) s is not the same state as s'." (Cleland 1991, p. 238)

That an event is productive means to involve concrete changes, of which the terminal states contain some sort of novelty. For one thing, a state with novelty should not be already there before the concrete change occurs. Novel thing should be a result of that change. For another, there are referring systems to estimate novelty. There is no absolute judgment for whether a state is novel or not, since a single state could be novel according to one referring system, but could be not, in correspondence to another.

There are two kinds of contextual determinants for novelty, and in further, for production. One is linguistic systems in everyday life and scientific practices. Linguistic systems in science especially refer to conceptual frameworks which are meaningful in particular fields. For instance, the novelty involved in DNA-RNA-protein synthesis is specified by conceptual frameworks of molecular biology. The other is material arrangement for intervention. It is necessary for determining productive property in science to consider particular experimental design, which aims to intervene and manipulate some natural process in the world. As mentioned before, neuroscientists have found that, under the condition of tetanic stimulation in a high frequency, the synaptic responses evoked by test stimulus will increase remarkably in a long run. To evaluate the productive property of the mechanism is to ascertain the novelty of the terminal state, "long term potential", which cannot be separated from the interventionist understanding of experimental design, such as tetanic stimulations in a high frequency and test stimulus.

---

[f]Humean theory of causation is reductivism, which claims non-causal properties are more fundamental than causal ones. However, Woodward (2003) argues that the interventionist theory should only be a characterization, rather than a reductive definition, of causation. The non-reductivists may believe that causal properties could be as the same fundamental as non-causal ones.

Concrete changes could always refer to verb-forms in ordinary or scientific languages. We even create new verb-forms in order to recognize newly found properties. Conceptual framework is usually transformed by new properties. In the era of scientific revolution, physicists have found that two electrons moving in the opposite direction with the same velocity could interact "at a distance". It seems to be a new kind of concrete change. And therefore we have a verb-form "quantum entanglement" which deeply alters the conceptual framework existing before scientific revolution.

Concrete changes could be also corresponded to material arrangements for intervention. For instance, molecular biologists recognize translation and transcription as new properties of production in the DNA-RNA-protein synthesis. And the concrete changes of translation and transcription could only be constructed from relations of manipulability, which means to change some variables of translation or transcription could influence other variables from protein-synthesis. It seems quite sure that the scientific understanding of concrete changes means possibilities for us to control things produced by altering things producing. That is why production is so significant.

6. Responses from Event-Property Theory

The event-property theory has provided a very different picture from verb-form theory. In terms of this new picture, some responses could be made to those objections in section 3 for verb-form theory. And I intend to show why its solutions are fruitful.

The first problem is that it seems necessary to make a distinction between those appropriate productive verb-form and other verbs merely superficial to characterizing production. According to event-property theory, whether a verb is appropriate for characterizing production depends upon whether it refers appropriately to some essential concrete changes. It is partly determined by linguistic systems, but also there is a role for concerns of intervention to take, since both are determinants for the novelty of concrete change.

Second, the claim for necessity in verb-form theory is seriously challenged by interventionists. How can we merely hold epistemic and methodological notions of necessity rather than an ontological one? For event-property theory, the verb-form theory only accepts linguistic systems as a contextual determinant for production; while it overlooks the other kind of determinant, concern of intervention which is fundamental to scientific experiments. If productive property configures itself by concerns of intervention, it seems unavoidable to locate manipulability relations on the ontological sense, as interpreted by interventionist theory.

It is worthwhile to analyze the debate in detail between verb-form theorists and interventionists. Event-property theorists would like to agree to the point that neither mechanism nor production could be independent from concerns of intervention. However, they would also point out, side with Machamer, that Woodward is wrong when he claims mechanism merely provides "spatial-temporal information" of manipulability relations. Causality contains both dependence and production. Dependence and production are actually coherent with and permeable to each other. Mechanism cannot be independent from concerns of intervention, while this does not imply any reduction from production to dependence. For productive property can be also influenced by something other than material arrangement for intervention, namely, linguistic systems for everyday or scientific practices.

In terms of ordinary or scientific language, there is significant difference between "kicking a ball" and "kicking a gearing device", or difference between "forming a covalent bond" and other bio-chemical activities. They are discriminated on kinds of concrete changes, which, according to the event-property theory, take essential roles in production. The difference seems unable to be illustrated by manipulability or causal dependence, which merely describe correlations among variables, but overlook activities of intermediate process. Interventionists may argue that those verb-forms could be accepted as existing predicates or n-adic relations. But then the concrete changes, referred to by verb-forms, would become black-box, neglecting thoroughly inferential relations in ordinary or scientific semantics. Only depends upon conceptual frameworks we are accustomed to in everyday life or scientific practices, can the kinds of concrete changes be clarified. Therefore it seems indispensable for linguistic system, as determinant of productive property, to guarantee irreducibility of production to dependence.

Would event-property theory also accept the irreducibility to canonical representation? It is a serious dilemma on characterizing production. I have ever argued that the verb-form theory is trapped in one horn, while Hall's reductivist analysis falls into the other. Event-property theory would prefer a reductivist analysis. As mentioned before, a determinable property, a "phase", seems essentially different from a determinate property, a "state". What represented directly by correlative variables are state changes, while phase changes may be directly illustrated by verb-forms. However, though it cannot be directly canonical representations of variables, phases could be indirectly defined on variables changes, as particular sets for types of state changes. Suppose there are two sets for state changes about temperature: {(1 °C, 20 °C), (30 °C, 12 °C), (11 °C, 39 °C), (70 °C, 72 °C)}, and {(100 °C, 101 °C), (1 °C, 4 °C), (5 °C, 100 °C), (20 °C, 200 °C) . . .}. Undoubtedly, most of us would tend to believe that there is no meaningful "phase change" represented by the first set, while the second one has expressed the rise of temperature, a significant phase change, referred to by "heating", for example.

Hall believes that the reduction of production to canonical form implies production in some sense "parallel to" dependence, which could be only discriminated from logical forms. Therefore the reduction would resolve the distinctiveness of production sooner or later, which is the second horn of the dilemma. Though it agrees to the reducibility, event-property theory does not hold Hall's implication. For instance, the rising of temperature could be defined as a particular kind of set for those concrete changes in which the value of terminal state become greater than original state. The reduction is possible rightly in virtue of concern for intervention or manipulability, a particular contextual determinant for productive property. The reduction mentioned above could be actually understood as, that the concrete change referred to by verb-form of temperature rising is equivalent to some sets for state changes such as {(100 °C, 101 °C), (1 °C, 4 °C), (5 °C, 100 °C), (20 °C, 200 °C) . . .}, in the very sense of intervening process of the world. However, the verb-form of "temperature-rising" has also other sense which is constructed by linguistic systems. To understand a notion presupposes some understanding of the inferential relations among conceptual framework in which the notion participates. For example, what temperature rising means depends upon its inferential relations to other relevant notions such as "lower", "fever", "melt", and so on. Those inferential relations could only be justified by semantics from ordinary or scientific languages, rather than any kinds of counterfactual dependence. In the sense of intervention and manipulability, we can believe in the reduction of productive property to canonical forms. Nevertheless, in the sense of semantics from linguistic systems, we can still defend some relatively independent status of production from causal dependence. Therefore event-property theory would not be trapped into the dilemma as what happens in Hall's analysis.

From the view of event-property theory, production cannot be reduced to causal dependence, while it is reducible to canonical representations. The reducibility to logical form does not imply that production and dependence could be distinguished only concerned to difference among logical forms. The essential difference exists in that production should also depend upon linguistic systems, which cannot be accommodated into concerns of interventions or manipulability. However, that is already enough for the purpose of scientific representation. For one thing, scientific representation attempts to specify what a particular notion of production is equivalent to in concerns of intervention or experimental design. That is to clarify the significance of concrete change for particular scientific practices. For another, those concerns of linguistic system seems not to be involved in scientific representation. To put forward or revise a theoretical model may actual extend or even alter the conceptual frameworks or scientific languages, in order to permit or refuse some particular inferential relation among notions, which also changes the inferential semantics of verb-forms.

Interventionists may agree to the event-property theory, or at least could accept some coherence between interventionist and event-property theories. Interventionist claims that we do not arrive at causal knowledge from any non-causal knowledge, but only infer new causation from some other causation we have ever known (Woodward 2003, pp. 104-5). The background knowledge for an experimental scientist usually contains not only knowledge about causal dependence from variable correlations, but also knowledge about production in the sense of its inferential semantics from conceptual frameworks.

7. Conclusion

Production is an essential notion for theory of mechanism, and to clarify its meaning brings about lots of difficulties. Event-property theory argues that production is a particular property of event which characterizes concrete changes. And it may provide a prospective solution for those difficulties. First, it can give a distinction between verb-forms which is appropriate for characterizing production and which is merely a superficial description. Second, it defends our intuition of causality that it is supposed to be ontological necessity besides methodological and epistemological ones. Finally, it points out a way out of the serious dilemma faced by both Machamer's verb-form theory and Hall's reductivist analysis. Therefore, the event- property theory seems to be a more appropriate approach for characterizing production. And it may provide an even cogent justification for theory of mechanism.

Acknowledgements

I would be grateful to the support from China National Social Sciences Fund (Grant No. 12CZX016), and the President Fund from University of Chinese Academy of Sciences.

References

1. D. Hausman, and J. Woodward. Independence, Invariance and the Causal Markov Condition. *British Journal for the Philosophy of Science* 50 (4): 521-583, 1999.
2. J. Woodward. Explanation and Invariance in the Special Sciences. *British Journal for the Philosophy of Science* 52 (2): 197-254, 2000.
3. S. Glennan. Mechanisms and The Nature of Causation. *Erkenntnis* 44: 49-71, 1996.
4. S. Glennan. Rethinking Mechanistic Explanation. *Philosophy of Science* 69: S342-S353, 2002.
5. P. Machamer, L. Darden, and C. Craver. Thinking about Mechanisms. *Philosophy of Science* 67 (1): 1-25, 2000.
6. P. Machamer. Activities and Causation: The Metaphysics and Epistemology of Mechanisms. *International Studies in the Philosophy of Science* 18 (1): 27-39, 2004.
7. S. Psillos. A Glimpse of the Secret Connexion: Harmonizing Mechanisms with Counterfactuals. *Perspectives on Science* 12: 288-319, 2004.
8. J. Woodward. What Is a Mechanism? A Counterfactual Account. *Philosophy of Science* 69: S366-S377, 2002.
9. J. Woodward. Mechanisms revisited. *Synthese* 183 (3): 409-427, 2011.
10. C. F. Craver, and L. Darden. "Discovering Mechanisms in Neurobiology." In *Theory and Method in the Neurosciences*, eds. P. Machamer, R. Grush and P. McLaughlin. Pittsburgh, PA.: University of Pittsburgh Press. 112-137, 2001.
11. W. Bechtel, and R. Richardson. *Discovering complexity : decomposition and localization as strategies in scientific research*. Cambridge, MA: The MIT Press. 2010.
12. B. Leuridan. Can Mechanisms Really Replace Laws of Nature? *Philosophy of Science* 77 (3): 317-340, 2010.
13. S. Psillos. "The idea of mechanism." In *Causality in the Sciences*, eds. Phyllis McKay Illari, Federica Russo and Jon Williamson. Oxford: Oxford University Press. 771-788, 2011.
14. G.E.M. Anscombe. "Causality and Determination." In *Metaphysics: Critical Concepts in Philosophy*, ed. M. Rea. New York: Routledge. 130-144, 2008.
15. N. Cartwright. "Causation: One Word, Many Things." In *Hunting Causes and Using Them*. Cambridge: Cambridge University Press. pp. 11-23, 2007.
16. N. Cartwright. *Nature's Capacities and Their Measurement*. Oxford: Clarendon Press, 1989.
17. R. Harré, and E. Madden. *Causal Powers: A theory of natural necessity*. Totowa, N.J.: Rowman and Littlefield. 1975.
18. S. Shoemaker. "Causality and Properties." In *Identity, Cause, and Mind: Philosophical Essays*. Oxford: Oxford University Press. 206-233, 2003.
19. C. F. Craver. *Explaining the Brain: mechanisms and the mosaic unity of neuroscience*. Oxford, Oxford University Press, 2007.
20. N. Hall. "Two Concepts of Causation." In *Causation and Counterfactuals*, eds. J. Collins, N. Hall and L. A. Paul. Cambridge, Mass.: The MIT Press. 225-276, 2004.

21. J. Bennett. "Event Causation: the counterfactual analysis." In *Causation*, eds. E. Sosa and M. Tooley. Oxford: Oxford University Press. 217-233, 1993.
22. D. Lewis. "Events." In *Philosophical Papers*. Oxford: Oxford University Press. 241-269, 1986.
23. L. B. Lombard. Causes, Enablers, and the Counterfactual Analysis. *Philosophical Studies* 59 (2): 195-211, 1990.
24. D. Davidson. *Essays on Actions and Events*. Oxford: Clarendon Press, 1980.
25. J. Kim. "Events as Property Exemplifications." In *Action Theory*, eds. M. Brand and D. Walton. Dordrecht: D. Reidel. 159-77, 1976.
26. L. B. Lombard. Events. *Canadian Journal of Philosophy*, 9 (3): 425-60, 1979.
27. C. Cleland. On The Individuation of Events. *Synthese* 86(2): 229-254, 1991.
28. J. Woodward. *Making Things Happen*. Oxford: Oxford University Press, 2003.

SOCIAL SCIENCE AND THE BAYESIAN PROBABILITY EXPLANATION MODEL

JIE YIN and LEI ZHAO

Research Center for Philosophy of Science and Technology, Shanxi University, Taiyuan 030006, China

C. G. Hempel, one of the logical empiricists, who builds up his probability explanation model by using the empiricist view of probability, this model encountered many difficulties in the scientific explanation in which Hempel is difficult to make a reasonable defense. Based on the bayesian probability theory, the Bayesian probability model provides an approach of a subjective probability explanation based on the subjective probability, using the subjectivist view of probability. On the one hand, this probability model establishes the epistemological status of the subject in the social science; On the other hand, it provides a feasible explanation model for the social scientific explanation, which has important methodological significance.

Keywords: Social Sciences; Scientific explanation; Bayesian probability explanation model; subjective probability.

According to Hempel, scientific explanation is to comment logically on the phenomenon with scientific laws, and to answer "why" questions proposed by scientists through the argument. The "deduction-nomological model" (i.e. D-N model) proposed by Hempel and Paul Oppenheim in 1948 and "Induction-Probabilistic model" (i.e. I-P model) proposed later were granted as classical explanation models in the field of the scientific philosophy, with measurable influence. As a basic programme of the logical empiricists, "unified science" was also a target that Hempel attempts to reach by logically rebuilding scientific explanation, for which Hempel generalizes these two explanation models to the whole field of the empirical sciences including the social sciences. But faced with a series of explanation dilemma, encountered by "I-P model" in the specific using, such as the requirement for high probability and the ambiguity of explanation, Hempel either avoided to reply or failed to make a reasonable defense. Just based on this background and Bayesian theory of Probability, this paper structures Bayesian probability explanation model under the horizon of subjective probability, which rebuild epistemological status of the exponent, aiming at providing a new methodological principle for social science which will guide people to recognize and understand the social sciences better.

1. Hempel's Dilemma on I-P Model and its Source

In his "I-P" explanation model, Hempel emphasized that in the Explanans there must be statistic laws, which, as the essential condition for explanation, played an important explanation role. Holding the naturalist view of science in the methodology, Hempel demonstrated that laws played a significant role in the social sciences similar with in the natural sciences. But his probability explanation model adopted the empiricist view of probability to interpret probability as the limit of the frequency, which was difficult to practice in the explanation of the social sciences. As to the social phenomena which were unavailable to obtain statistic laws through the empirical observation, it was unlikely for "I-P" explanation model to resort to statistic laws to give a reasonable explanation. So the initial question that Hempel faced with during applying the I-P model to the explanation in the social sciences was how to consider the scientific explanation excluding statistic laws. The source of this dilemma derived from a basic philosophical proposal that the logical empiricism rejected metaphysics However, it was also the inherent requirement of logical empiricism's programme to adopt the empiricist view of probability.

1.1. *Classical model of scientific explanation: I-P model and its dilemma*

The essence of Hempel's probabilistic explanation model is inductive arguments. The fundamental schema can be represented as follows:

$$\begin{array}{c} (\mathrm{I-S})\,\mathrm{Pr}\,(\mathrm{G}|\mathrm{F}) = \mathrm{r} \\ \mathrm{F\,i} \\ \hline\hline \mathrm{G\,i} \end{array} \quad [\mathrm{r}]$$

"The double line separating the premises from the conclusion in this schema signifies an inductive argument; the bracketed r is the degree of inductive support given to the conclusion by the premises. The first premise is taken to be a statistical law."[1] In I-P model the premises were regarded as "Explanans" which consist of statistic law and antecedent conditions, and conclusion as "Explanandum". F and G express category, and i expresses a single event. Taking this statistical law $\mathrm{Pr}(\mathrm{G}|\mathrm{F}) = \mathrm{r}$ as premise, this explanation argument point out that the single event i is a case in F so as to explain the particular fact that i is G. In scheme the relative frequency that G occur was r in the case of F.

Aiming at the I-P model features above mentioned, One comes up with lots of counter-examples, in which ambiguity of statistic explanation is one of annoyance problems Hempel must face to. For instants, streptococcus counter-example — John infects streptococcus, but he recovers by taking Penicillin. According to I-P model

[1]Hempel, C. G. 1965: Aspects of scientific explanation. In Aspects of Scientific Explanation and Other Essays in the Philosophy of Science (New York: Free Press), p. 390.

as follow:

$$\frac{\begin{array}{c}P(R(x), S(x) \wedge P(x)) = 0.95 \\ S(j) \wedge P(j)\end{array}}{R(j)} \; 0.95$$

The pattern above means that the probability of recovery R, as someone infects streptococcus S, is highly 95% after taking the penicillin, particularly, John j that infects streptococcus S therefore had recovered when he took the penicillin. However, the probability of failing to recover is almost 65% if John happens to infect F streptococcus which has resistance to Penicillin after taking it. This case can be explained as follow:

$$\frac{\begin{array}{c}P(\neg R(x), S(x) \wedge P(x) \wedge F(x)) = 0.65 \\ S(j) \wedge P(j) \wedge F(j)\end{array}}{\neg R(j)} \; 0.65$$

Comparing two patterns above mentioned, we can learn about whether John is recovery or not, I-P model all offers rightly explanation for two different case, that is, the two propositions consisted of two different true propositions respectively provide better high probability.

Hempel's probabilistic explanation, in which the laws used is statistic laws, only assert that some kind of phenomena or events appear in a certain way of probability, which have the nature of induction, and that also merely provide the partial explanation for some fact. The conclusion is not deduced from this sort of scientific explanations on the basis of the universal laws, but supported by a certain degree of probability on the basis of the statistical laws. That means the relationship between Explanans and Explanandum is not universally necessary but probabilistic. That is, the conclusion Gi is not deduced necessarily from conjunction of statistical law $Pr(G|F) = r$ and the antecedent conditions, but deduced probably.

1.2. *Analysis of the root of dilemma in I-P model*

According to J. H. Fetzer, Hempel's probability model failed in his probabilistic views, because he accepted frequency interpretation of probability which is counted as the empiricist view of probability. Because of that, this directly leads to two dilemmas of I-P model in scientific explanations, which are high probabilistic request and ambiguous problem of scientific explanations. All of these are caused by I-P model. Specifically, "induction" means there is probabilistic relation between Explanans and Explanandum.

The analysis of the empiricist view of probability that cause the dilemmas of I-P model are as follow.

Hempel's scientific explanation model based on probabilistic statistic laws provide probabilistic inductive support for Explanandum. In the repeated trail principle Statistic laws are probabilistic empirical statistic laws by empirical observations. Hempel argues that statistic laws are lawlike statements with higher frequency and are different from accidental generalization. Empiricists interpret probability by use of frequency, so frequent interpretations of probability can also thought to be empiricist probabilistic explanations. In classic statistics, "The frequentists approached statistical stability from a philosophically empiricist point of view."[2] They claimed that probability is frequency, that is to say, probability is the limit of a long-run relative frequency. For example, in n times repeated trials, if we note m as the times of event A that occurred, then the probability of event A is:

$$P(A) = \lim_{n \to \infty} \frac{m}{n}$$

this probability means frequency that event A occurred. In the other words, the frequentists define probability as the limit of frequency that event occurred in infinitely repeated trials. "Frequentists propose to apply their concept of probability to 'statistical' phenomena."[3] It can be perceived that the methodological principles of the probability is essentially empirical, and the probability of empiricism define the probability by depending on people's experiences that one have observed the frequency of some event occurred.

The empiricist view of probability was problematic, this also brought about explanation dilemmas that Hempel's scientific explanations model was unable to pass for its application in the social sciences. Firstly, as for the social phenomena that it's unavailable to obtain adequate sample through the empirical observation how can we find statistic laws with higher frequency? Secondly, how can we regard the sort of explanation in the social sciences which fail to contain the statistic laws? That is to say, in the research of the social sciences, how can we provide reasonable scientific explanations for particular fact that it is unavailable to discover statistic laws. Thirdly, the free will in the social sciences cause the complexity and uniqueness, whereas the empiricist view of probability entirely exclude epistemic subjective factors, therefore there is a problem, that is, how to include the subjective factors into the logical analysis of the model?

In this paper, we suggest that the source of this dilemma, which Hempel's probability model encountered in the social sciences, derives from a basic philosophical proposal of logical empiricism — logical empiricist focus only on the analysis of the logical form in order to reject the metaphysics; and excluded the epistemological status of the subject that served as explanations, moreover, it was also the inherent requirement of logical empiricism's programme to adopt the the empiricist view of probability. Therefore, this article also rightly realizes the limitation of the empiricism in the social sciences, and introduces another view of probability, the subjectivist view of probability, and then based on that view constructs a

[2]Michael Strevens. Bigger than Chaos Understanding Complexity through Probability[M]. Harvard University Press, 2003, p. 9.

[3]W. H. Newton-Smith. A Companion to the Philosophy of Science[M]. Blackwell Publishers Ltd, 2000, p. 367.

new model of probabilistic explanation — Bayesian probabilistic explanation model. The construction of this new model reestablishes the core status of the exponent in the explanation process. At the same time it contains the fascination of the model of Hempel's probabilistic explanation and avoids dilemma on I-P model, showing strongly explanatory power and vitality.

2. The Introduction of Bayesian Probability Explanation Model in the Social Sciences

Based on the analysis above, in order to solve the dilemma of Hempel's probability explanation model in the social sciences, we just need to relocate the epistemological status of the exponent as the explanatory subject, and to introduce the background knowledge or cognitive states of the explanatory subject as well as the exponent's degree of beliefs for the relevant factors into explanatory factors of social sciences. The probability explanation model based on Bayesian method constructs a feasible platform for this introduction. Based on bayes theorem, Bayesian method has developed to solve the statistic problems, which could solve incompleteness and uncertainty of the explanation as well as possible. Therefore, this essay suggests that it is feasible to describe the exponent's degree of beliefs in the explanation of social sciences by use of the Bayesian method, and then we call this explanation model as Bayesian probability explanation model (Simplified as B-P model). For one thing, this model is able to introduce the epistemological status of the exponent into the explanation; for the other, it effectively eliminates the difference caused by different explanations given by different exponents to the same empirical observation, so it has great methodological significance.

2.1. *The theoretical source of B-P model*

The subjectivist view of probability is adopted by The B-P model, in which "probability is usually interpreted in subjective terms a formal, mathematically rigorous characterization of beliefs."[4] The subjectivist interprets the probability as the degree of belief of rational agent, namely, the degree of an agent believing that some event occurs. The problem of the single event assigning probable values involves the degree of belief of agent, therefore the phenomena of the belief of degree cause the subjectivism. The subjectivist theory of probability is concerned with the measurement problem of the degree of belief, so it belongs to one of the psychological social or human sciences. The subjective probability is also known as "personal probability", and this doctrine of probability rightly is Bayesianism or "personalism". In the framework of bayes theory, "It involves various assumptions: that beliefs differ

[4]Simon Jackman. Bayesian analysis for the Social Sciences[M]. A John Wiley and Sons, Ltd, Publication, 2009, p. 3.

in strength, not just qualitatively but quantitatively, that the numerical functions specifying these quantities for rational agents satisfy the axioms of the probability calculus, and that values of these functions are sufficiently well connected to the existing employment of the term 'probability' to be deemed probabilities."[5] That is to say, the strength of belief can be measured by the numerical values. F. P. Ramsey, who is a subjectivist, writer by oneself that "It is not enough to measure probability; in order to apportion correctly our belief to the probability we must also be able to measure our belief."[6] The traditional method of measuring belief just is of proposing bet. Bayesians consider that, for a rational agent, we can determine subjective probabilities of particular propositions by means of determining the odds that the person would be willing to accept in a bet on that proposition.

The subjectivist interpret probability as the degree of belief, which is interpreted as fair betting-quotient. This practice have heavily supported by the famous theorem of Dutch Book that "if the P1, P2 ... are a set of betting quotients on hypotheses H1, H2, ..., then if the Pj do not satisfy the probability axioms, there exists a betting strategy and a set of stakes such that whoever follows this betting strategy will lose a finite sum whatever the truth values of the hypotheses turn out to be."[7] Sometimes this theorem is known as the Ramsey — de Finetti Theorem, in which the betting quotients is the ratio between the bet a bettor could be willing to make and all bets. The significance of theorem of Dutch Book is that it is reasonable to interpret the probability as the degree of belief. Particularly if the results of Dutch Book extend to the case of conditional probabilities, the better do not update his subjective beliefs according to new information in consistent with the probability axioms, and then he will be in a losing station. That is, when probability may be subjective, Bayes theorem governs people how to update subjective beliefs rationally.

The theory of Bayesian probability regards Bayes theorem as the model that agent update prior probability in light of empirical facts, and this model, in which it bases on background knowledge of the exponent, assigns the prior probability to the probability that specific facts occurred, but the prior probability is not without changing. With agent continually acquiring empirical facts, the prior probability is updated or adjusted by the rule of probability calculation, which is Bayes theorem, thereby, the initial difference of the prior probability is "eliminated". This process makes the subjectivity and personality of heterogeneous subjects for the prior probability acquire objectification or publicity, therefore plays a role in "studying from experience" as well as possible.

[5]W. H. Newton-Smith. A Companion to the Philosophy of Science[M]. Blackwell Publishers Ltd, 2000, p. 361.

[6]Antony Eagle. Philosophy of probability contemporary readings[M]. London and New York, Routledge, 2011. p. 53.

[7]Howson, Colin and Peter Urbach. Scientific Reasoning: the Bayesian Approach[M]. Second ed. Chicago: Open Court, 1993, p. 79.

2.2. *The construction and application of B-P model*

The key of solving the dilemma of Hempel's probability model in the social sciences is that we need to introduce the background knowledge or cognitive states of the explanatory subject as well as the exponent's degree of belief for the relevant factors into explanatory factors of social sciences, that is to say, we must relocate the epistemological status of the exponent as the explanatory subject. In the certain explanation process, the degree of beliefs possessed by the exponent for the relevant factors concerning both Explanans and Explanandum are changed with the explanatory context changing, therefore, the exponent's degree of beliefs for the relevant factors have the characteristics of incompleteness and uncertainty, and bayes method could solve incompleteness and uncertainty of the explanation as well as possible. For this reason, based on the characteristics of the theory of bayes probability, we can construct a standard method of psychology for the explanation of the social sciences, "many believe that the basic principle of such a model involve a probability function. This function is a so-called personal probability, a function specific to the agent which captures the degrees to which he believes the various propositions he can conceive."[8] This explanation model can be stated that the exponent assigns a prior probability to Explanans basing on his background knowledge, and then with the context changing, the exponent acquires new empirical evidences, and those evidences update the exponent's prior probability to acquire the posterior through the rule of the conditional probability determinated by the bayes theorem, that is $p(\theta/\mathbf{y}) = \frac{p(\mathbf{y}/\theta)p(\theta)}{\int p(\mathbf{y}/\theta)p(\theta)d\theta}$, thereby the exponent make rational explanation of probability in the different context. Among the conditional probability, it means that θ indicates the uncertainty of the exponent against a proposition or event; $p(\theta)$ is called the prior probability of θ, and $p(\mathbf{y}/\theta)$ shows probability function, that is to say, likelihood function, after exponent acquiring empirical facts; $p(\theta/\mathbf{y})$ indicates the posterior probability of θ.

In the following section we will describe the application of the B-P model against the counterexample of the I-P model that the treatment effect of streptococcal infection for the penethamate.

In this counterexample, first of all, according to exponent's background knowledge that may be his experience, he or she (e.g. a medical expert) assigns higher prior probability, say, the probability of recovery to be more than 95%, to Explanans "which is a proposition that taking the penethamate can effectively cure the streptococcal infection", so John recovers by he taking the penethamate. In addition, this medical expert also research for many years concerning influencing factor that affect treatment effect, such as concurrent infection in patient, the immunoreaction of patient by oneself and so on. Aimed at John taking the penethamate but no recovering, this medical expert makes a examination against John's disease and concludes that John infects a kind of streptococcus which has resistance to Penicillin, and then the

[8] W. H. Newton-Smith. A Companion to the Philosophy of Science[M]. Blackwell Publishers Ltd. 2000, p. 360.

exponent updates the antecedent prior probability after acquiring that empirical evidence, in which the logical law as updating is Bayes theorem, thereby the exponent obtains a posterior probability about recovering for John, namely, the probability of recovery drops to 35% below, so the exponent makes a explanation that may be a certain rationality for the counterexample above throughout that process of explanation above. Therefore, we find that it is not the statistic laws that the B-P model resort to, and this model assigns a prior probability (95%) to treatment effect in light of the exponent's background knowledge or cognitive status, in addition, with the explanatory context changing, this prior probability is updated by Bayes theorem in light of the new empirical fact (i.e. a sort of streptococcus resisted to Penicillin), and the exponent obtains the posterior probability (35%). Therefore, the B-P model is extremely applicative for the social phenomena that are difficult to find the laws.

To sum up, the probability explanation of specific events in the explanation of the social sciences is relative to the exponent's background knowledge, and the B-P model introduces the exponent's degree of beliefs for the relevant factors into explanatory factors of sciences. For one thing, this explanation model, which has dynamic epistemological significance, re-establishes the exponent's epistemological status in the process of explanation, and so gives an access to solve the problem of the ambiguity of explanation; for another, because of the characteristic of peculiarity and non-repeatability and confirmatory with difficulty, and so forth, arming at those characteristic the B-P model also makes greatly probability explanation for social phenomena that are difficult to find.

3. The Significance of the Introduction of Bayesian Probability Explanation Model in the Social Sciences

Applied empirical tools and logical technology, Hempel's model defines the scientific explanation as finding objective relation between Explanans and Explanandum, not the subjective beliefs associated with cognitive subject. But the exponent accordingly gives different explanation as changing of context, and both the essence of scientific explanation and the exponent's the background knowledge as well as cognitive state are inseparable, therefore, it is a necessary premise for the scientific explanation that subjective element, say, exponent, is introduced into the element of scientific explanation. In the form of prior probability B-P model integrates the exponent's the background knowledge into the logical analysis of explanation, and by the Bayes theorem the exponent updates or adjusts the prior probability in different context, so this process embodies a cognitive dynamics.

3.1. *B-P model: A dynamic epistemological model*

The function of Hempel's probability explanation model lies in confirming of logical relation between Explanans and Explanandum, and this model devotes to logical argument of the explanation of single event, in which the statistic laws play a very important role. It is generally admitted that in fact the ability of explanation for the scientific explanation is undertaked by the laws. The statistic laws play a associated role, and they associate specific event mentioned in antecedent condition in Explanans with specific event mentioned in Explanandum. Hempel's probability explanation model adopts the empiricist view of probability in order to reject the metaphysics, and entirely excludes the epistemological status of explanatory subject, however, this practice that the explanatory model of nature sciences applies to social sciences heavily ignore the peculiarity which the social sciences possess. Just in the form of subjective probability Bayesian probabilistic model integrates the exponent's the background knowledge or cognitive status into the explanation, and avoids the dilemma of Hempel's probabilistic model, and then provides a new explanatory model that is suited for the social sciences.

The construct of B-P model derives from the Bayesian probability theory, Baysians consider that the probability of an event is a personal belief that the researcher provide for the probability of this event in light of his background knowledge. Such probability is called subjective probability. Consequently, "Bayesian probability statements are thus about states of mind over states of world, and not about states of the world per se,"[9] the probability is interpreted as a subjective degree of belief by the subjectivist, this interpretation embody the individual level of uncertainty to the researcher in the occurrence of an event. The one of the major effects of Bayesianism is that there is no necessary for researcher to specify or conceive of some relevant sequence of repetitions of the event, as the frequentist does, and this advantage enables the probability of the "one-off" sorts of events to be estimated consistently by the researcher. "Prior information is formalized in the Bayesian framework and this prior information can be subjective in the sense that the researcher's experience, intuition, and theoretical ideas are included,"[10] such as the medical expert's medical practice for many years. In the Bayesian philosophy, the probability assigned to the occurrence of any events depends upon the individual whose uncertainty is taken into account and upon the state of background information for the individual. Therefore, according to the view of subjectivism, there no exist the hypothesis that any uncertain events have a sole value of probability, and arming to the probability of the same event Bayesians allow the different individual to possess different subjective degree of beliefs.

In the Bayesian probability theory, the logical calculation law from the prior probability to the posterior probability is bayes theorem, and the posterior probability also serve as the prior probability in the next calculation, consequently, doing this calculation repeatedly can be eliminated the initial difference of exponent's prior probability. In this typically updating sense, B-P model is a kind of

[9]Simon Jackman. Bayesian analysis for the Social Sciences[M]. A John Wiley and Sons, Ltd, Publication, 2009, p. 7.
[10]Gill, Jeff. Bayesian Methods: A Social and Behavioral Sciences Approach[M]. The CRC Press, 2002, p. 5.

epistemological model. This model have two major characteristic, in which one is that the probability is interpreted as a person's degree of belief, that is to say, the exponent's subjective probability; the other is that bayes theorem is regarded as a way that can change degree of belief in light of empirical facts. In the other word, this model introduces the exponent's irrational factors into the scientific explanation by means of the strictly logical bayes laws, namely, bayes theorem, and highlights the core status of the exponent as the explanatory subject, and also allows the exponent to consider such all available information, for example the former researching conclusion, and then its dynamism in the explanatory procedure is embodied in bayes theorem that has repairability. B-P model emphasizes the background knowledge of exponent, while that of exponent is an important basis on which the exponent determines prior probability. This model at least paves the way for the relation between the facts and beliefs, in which the facts is about the world's facts and the beliefs is the exponent's beliefs as cognitive subject.

All in all, the function of the dynamic epistemology of B-P model makes explanation of social sciences much more rationality. This model bases on the concept of subjective probability, however, this concept does not depends upon the statistic laws at all in the process of explanation. So, it is more universality than I-P explanation model. In closing this section, we summarize the viewpoint of the B-P explanation model by contrasting it with the I-P explanation model.

Table 1. Comparison between the B-P explanation model and I-P explanation model

	B-P explanation model	**I-P explanation model**
What kind of view of probability to being adopted?	The view of subjectivist probability.	The view of empirical probability.
What is probability?	The degree of beliefs which obey the axioms of the probability.	The limit of a long-run, relative frequency.
How is probability measured?	By betting procedures (as betting quotients).	By repeating trials.
What is the explanatory ways?	The condition probability confirmed by bayes theorem.	The explanation, namely, demonstration.
What is nature of the model?	A dynamic epistemological model.	The purely formal model.
Whether they require the laws or not?	Presence or absence can be	It must appeal to the laws.
Whether they contain the subjectival elements of exponent or not?	Yes	No

3.2. *Defending on epistemological status of the explanatory subject in the explanation for the social sciences*

Severing as a respond for post-kantianism in the nineteen century, "The fundamental tenet of logical empiricism is that the warrant for all scientific knowledge rests upon empirical evidence in-conjunction with logic, where logic is taken to include induction or confirmation, as well as mathematics and formal logic."[11] Regarding the metaphysics and theology to be rejected as basic programme by logical empiricism, they make a stand against speculative metaphysics, who proposal empiricism in the epistemology, who emphasize on the importance of the modern logic. In this background, Hempel, who is one of members of the logical empiricism, adopt both the logic and experience as the tools of philosophical analysis, and then, by the completely formal ways, make a study of the scientific explanation, meanwhile, also promote the scientific explanation to become one of the themes in the philosophy of science. "The problem of explanation for logical empiricists was to find some conditions of explanation which insure the objective relevance of the explanans to the explanandum. They needed a relationship which made explanatory relevance a matter of objective relations between statements and not the subjective beliefs about relevance of less-than-omniscient cognitive agents."[12] The explanation model that constructed by Hempel as a logical empiricist is not eager to consider the exponent' belief and interest with respect to scientific explanation, but pays much closer attention to both the logical construction and relative demonstration between Explanans and Explanandum.

The theory of Hempel's scientific explanation emphasizes on logical reconstruction for the scientific explanation. He divides the sciences into the empirical sciences and non-empirical sciences, respectively, in which the empirical sciences contain the natural sciences and the social sciences. As far as he is concerned, the same explanation model can be applied to both the natural sciences and the social sciences, therefore he devoted the explanation model to apply to the social sciences after he puts forward the formal explanation model. He points out that the explanation of social sciences are similar to the logical construct of natural sciences like physics, and then the human actions are explained by finding the proper laws just as the natural sciences. However, for one thing, the objects of study in the kind of the social sciences have a common characteristic — the subjectivity in the study of social sciences, "The subjects of social science are themselves epistemic agents, who can be influenced by their own beliefs about the generalizations, the input, and the output of social theory."[13] The subjectivity of the study subject make the heterogeneity

[11]W. H. Newton-Smith. A Companion to the Philosophy of Science[M]. Blackwell Publishers Ltd, 2000, p. 233.

[12]Alexander Rosenberg. Philosophy of Science A contemporary introduction[M]. London and New York, Routledge, 2001, p. 27.

[13]W. H. Newton-Smith. A Companion to the Philosophy of Science[M]. Blackwell Publishers Ltd, 2000, p. 457.

in the social sciences particularly prominent, which is greatly different with the natural sciences in which the basic theoretical premise and hypothesis are subject's homogeneity, so social science are more complex than natural science. I-P model constructed by Hempel who is a logical empiricism, entirely excludes the explanatory cognitive subject, but the difference in the scientific explanation substantially is a difference among the background knowledge or theoretical systems of the exponent, and this kind of difference is extremely significant in the social sciences. Therefore the dilemma of I-P model applying to the social sciences is inevitable. It can be perceived that the epistemological status embodied by the background knowledge or cognitive state is extremely important in the explanatory factors in the explanation of the social sciences.

For another, some social phenomena, to some extent, fail to be observed and put to test repeatedly. Because of the one-dimensionality or nonrepeatability or nonverifiability represented by the social sciences, the empirical laws summarized by the exponent is difficult to apply to the same kinds of events another time. Because of that, people propose a challenge for the basic of objectivity of the social sciences. Hempel regards these characteristic in the social sciences as incompleteness possessed by itself, and thoroughly applies the research method into the research of social sciences, however, because "Social phenomena result from the activities of human beings, and human beings are agents whose actions are directed by their beliefs, goals, meanings, values, prohibitions, and scruples,"[14] and it is difficult to encounter the explanation that all the people can accept. So, the heterogeneity of the cognitive subject in the social sciences make the researcher to find few laws, and the explanation that fail to apply to laws only depend upon cognitive or theoretical framework of Scientist, for instance, "Remmer identifies economic crisis as a source of electoral volatility in Latin American countries in the 1980s;"[15] "Golden finds that union wage moderation in industrialized democracies results from coordinated collective bargaining."[16] Therefore, in the great majority of explanation of the social sciences, we take the explanation to be sought and the subjective beliefs to be provided by the exponent as a basic of factors of scientific explanation. As to such problems of the social sciences, I-P model based on statistic laws fail to resort to laws to carry out explanation.

To sum up, a really significant scientific explanation is not only concerned with formally logical argument, but also the exponent's background knowledge or beliefs. Hence, the explanation of the social sciences urgently needs the effective intervention of the explanatory factors of subject, containing background knowledge or cognitive states or cognitive conditions and so on, and then the construction and applying of

---

[14]Daniel Little. Varieties of Social Explanation An Introduction to the Philosophy of Social Science. Westview Press, 1991, p. 39.

[15]Remmer, Karen L. The Political Impact of Economic Crisis in Latin America in the 1980s. American Political Science Review, 1991, 85:777–800.

[16]Golden, Miriam. The Dynamics of Trade Unionism and National Economic Performance. American Political Science Review, 1993, 87:439–54.

B-P model in the social sciences make the subjectivity factors effectively integrate into the scientific explanation factors, and reassign the exponent's epistemological status.

4. Conclusion

The root of dilemma of Hempel's I-P explanation model is that it rejects metaphysics, and only focuses on logical argument of the explanation, and fails to pay close attention to the epistemological status of exponent as cognitive subject, and adopts the logic and experience, in its explanatory process, as a type of tools of philosophy analysis, and insists on logical reconstructed programme, namely, the explanation is argument. The difference in the standard of the scientific explanation substantially is the difference among exponent's background knowledge or cognitive state, and that difference is extremely significant in the social sciences. Based on the Bayesian probability theory, B-P explanation model regards the subjective interpretation of probability as the degree of rational beliefs, and reconstructs the epistemological status of the exponent. We introduce the degree of beliefs concerning relative factors in the different explanation context into the scientific explanation factors via this model. This model constructs a sort of explanatory model that is more effective or reasonable for the social sciences, which has important methodological significance.

References

1. Colin, H. and Urbach, P. (1993). *Scientific Reasoning: the Bayesian Approach. Second ed. Chicago:*, Open Court.
2. Eagle. A. (2011). *Philosophy of probability contemporary readings.*, London and New York, Routledge.
3. Gill. J. (2002). *Bayesian Methods: A Social and Behavioral Sciences Approach*, The CRC Press.
4. Golden, M. (1993). "The Dynamics of Trade Unionism and National Economic Performance", in American Political Science Review, 87.439-54.
5. Hempel, C. G. (1965). "Aspects of Scientific explanation". In *Aspects of Scientific Explanation and Other Essasy in the Philosophy of Science.* New York : Free Press
6. ______ (1996). *Philosophy of Natural Science.* Prentice-Hall.
7. Jackman, S. (2009). *Bayesian analysis for the Social Sciences.* A John Wiley and Sons Ltd., Publication.
8. Little, D. (1991). Varieties of Social Explanation An Introduction to the Philosophy of Social Science, Westview Press.
9. Newton-Smith, W.H.(2000). *A Companion to the Philosophy of Science*, Blackwell Publishers Ltd.
10. Remmer, K.L. (1991). "The Political Impact of Economic Crisis in Latin America in the 1980s" in american Political Science Review, 85:777-800.
11. Rosenberg, A. (2001). Philosophy of Science A contemporary introduction. London and New York, Routledge.
12. Strevens, M. (2003). Bigger than Chaos Understanding Complexity through Probability. Harvard University Press.

CHOICE OF UNITS AND THE CAUSAL MARKOV CONDITION

JIJI ZHANG

Department of Philosophy, Lingnan University
Tuen Mun, NT, Hong Kong
jijizhang@ln.edu.hk

PETER SPIRTES

Department of Philosophy, Carnegie Mellon University
Pittsburgh, PA, USA
ps7z@andrew.cmu.edu

Elliott Sober's well-known challenge to the principle of the common cause — and to its generalization, the causal Markov condition — appeals to the apparent positive correlation between two causally unconnected quantities: Venetian sea levels and British bread prices. In this paper we examine Kevin Hoover's and Daniel Steel's opposite evaluations of Sober's case. We argue that the difference in their assessments results from a difference in their choice of units and populations for statistical modeling. Our analysis suggests yet another diagnosis of Sober's counterexample: the failure of the causal Markov condition in the population chosen by Sober and Steel is due to the presence of causal relations that hold between the relevant properties *across* units. Such inter-unit causation is left unrepresented in causal models congenial to statistical analysis, because statistics does not deal with inter-unit relationships once the units are fixed. Accordingly, the causal Markov condition is formulated in terms of causal structures that depict intra-unit causal relations only. It is therefore worth highlighting a methodological principle for causal inference: the units should be so chosen that they do not interfere with each other, a principle that, fortunately, is often observed in practice.

1. Introduction

For people who take causality and probability seriously, there seems to be little controversy that probabilistic dependence has something to do with causal relations. A precise formulation of the connection, however, has been subject to some debate. The formulation is known as the *causal Markov condition* (CMC), a revival and generalization of the *principle of the common cause*, which is traceable at least to John S. Mill (1843), and made influential by Hans Reichenbach (1956). The condition specifies the exact manner in which probabilistic independence relations are entailed by a given causal structure, and serves as a springboard for much of the work on statistical inference of causal relations and quantities (e.g., Spirtes et al. 2000, Pearl 2009).

The CMC states that given a causally sufficient set of variables, every variable in the set is probabilistically independent of its non-effects (i.e., variables that are not its effects) conditional on its direct causes. The requirement of causal sufficiency is important; some apparent counterexamples to the CMC feature systems that are not causally sufficient (Spirtes et al. 2000, pp. 33-35). Roughly, a set of variables is causally sufficient if no direct common cause of any two variables in the set is left out.

A special case of the CMC obtains when two variables form a causally sufficient set — that is, they do not share a common cause — and neither of them causally influences the other. The CMC then implies that the two variables are probabilistically independent. In other words, if two variables are not probabilistically independent, then either one of them is a cause of the other, or there is a common cause (that renders the two conditionally independent). This of course is a familiar formulation of the principle of the common cause.

Elliott Sober's well-known challenge to the principle of the common cause involves two apparently correlated quantities that are arguably causally unrelated: Venetian sea levels and British bread prices (Sober 1987). Two opposite evaluations of this case have come to our attention. Kevin Hoover (2003) argues that Sober's claim is based on an incorrect measure of probabilistic dependence in the population, and the counterexample disappears if the right measure is used. By contrast, Daniel Steel (2003) argues, based on a theorem about mixing populations, that Sober's case is a genuine counterexample to the principle of the common cause[a], because probabilistic dependence is provably present in the population.

In our view, the opposition between their conclusions regarding the legitimacy of the counterexample results from a difference in their choices of units and population for the causal analysis. Relative to their respective choices, both authors present sound diagnoses of the problem. We wish, however, to highlight a methodological principle that favors Hoover's choice of units in the Venice-Britain scenario. The principle is that in defining a population for the purpose of causal analysis, one should choose "units" in a way that avoids, as much as possible, causal interference across the chosen units. A simple rationale behind this principle is that in the currently standard language for statistical modeling of causation, only intra-unit causation, but not inter-unit causation, can be represented. Accordingly, as we will argue, the CMC (as well as any other bridge principle one may propose to link causality with probability) is formulated in terms of the structure of intra-unit causal relations. The presence of the unrepresented inter-unit causation may contribute to a probabilistic dependency in the population that cannot be accounted for by the intra-unit causal structure — if so, the CMC fails.

[a]Although Steel approves of the counterexample, he objects to Sober's proposal of subsuming the principle of the common cause into the principle of likelihood. For the present purpose, we will focus on his defense of Sober's case.

2. Hoover and Steel on Sober

Sober (1987) cites the relationship between the sea levels (SL) in Venice and the bread prices (BP) in Britain as a violation of the principle of the common cause. Both quantities have seen a steady monotonic increase over time. If we calculate the correlation (say, Pearson's correlation coefficient) between the two quantities using the values from different periods, it will turn out strongly positive.[b] Moreover, this positive correlation will not diminish as more and more periods are included in the calculation, which suggests genuine probabilistic dependence in the population. However, SL does not influence BP and BP does not influence SL. It is also very unlikely that there is a common cause that fully accounts for this probabilistic dependence, or so we can assume for the sake of the argument. Thus the principle of the common cause is patently violated.

Several diagnoses of the case have since been given (e.g., Papineau 1992; Hausman and Woodward 1999; Spirtes et al. 2000). We will focus on two more recent responses to motivate the diagnosis we favor. Before analyzing their arguments, it is helpful to first clarify what we mean by the choice of units. Statistical inference is usually conceived as reasoning about a population based on samples. By "units" we mean elements in an (often hypothetical) population from which samples are supposed to be drawn. Each unit is an (often abstract) entity that instantiates or realizes a set of variables of interest. In general there is little restriction on what a unit could be. Depending on the context of analysis, a unit could be a single concrete entity such as a person or a physical system, a social organization such as a family or a firm, or even a geographical region at a historical time, say, Europe in 1990.

A typical kind of change in the choice of units is to systematically combine current units to form new, bigger units, or conversely, to decompose the current units into smaller ones. For example, an analyst may shift from a population of individual persons to a population of couples by combining the original units into bigger ones, i.e., pairs of persons. Such a change of units is usually accompanied by a redefinition of variables. The variable Age, for example, may be redefined as $AverageAge$ given the new choice of units. Moreover, there can be two Age variables given the new units, Age_1 and Age_2, the probabilistic relationship between which is now amenable to statistical analysis; such an analysis is not possible given the original choice of units. On the other hand, if one is only interested in the relationship between, say, Age and $Weight$, it is probably a better idea to hold on to the original choice of units for the ease of sampling. The point is that units are chosen for the sake of

---

[b]The mathematical aspects of such correlations were explored in detail by G. Udny Yule (1926), who regarded such sample correlations "nonsensical" (his leading example involved the proportion of Church of England marriages and the mortality rate, both of which had a general, though not strictly monotone, falling trend during 1866-1911). Yule demonstrated that for some classes of time series, the usual significance test of correlation is inapplicable or at least very inefficient. This point is closely related to Hoover's diagnosis of the situation. Interestingly, Steel's argument, as we will see, was also anticipated in an even earlier paper by Yule (1903).

analysis on largely pragmatic grounds.

Back to Sober's case. Hoover (2003) points out that the scenario involves two *non-stationary* time series, for which the usual sample correlation is not a reliable indicator of probabilistic dependence or independence. To judge whether two (integrated) non-stationary series are probabilistically (linearly) dependent, for instance, the criterion of *cointegration*[c] should instead be used. Hoover concludes that since the Venetian sea levels and the British bread prices are not cointegrated, they do not exemplify the kind of probabilistic dependence relevant to the principle of the common cause. Thus Sober's scenario does not make a genuine counterexample.

We find Hoover's diagnosis persuasive.[d] It is important to notice that Hoover talks about probabilistic dependence between two stochastic *processes* rather than two ordinary random variables. A stochastic process, strictly speaking, is constituted by a set of distinct random variables whose realizations are ordered in time. Thus, in the analysis of the relationship between the SL process and the BP process, a unit in the underlying (hypothetical) population is a realization of the two entire processes rather than an instantiation of the two quantities in a single period. Relative to this hypothetical population, the SL process and the BP process are probabilistically independent. For the set of SL variables $\{SL_t\}$ and the set of BP variables $\{BP_t\}$ are probabilistically independent. In practise, such a choice of units of course poses challenging statistical issues — there is usually only one realization of a segment of the series available. That is where such conditions as stationarity can be extremely helpful in amplifying sample size and justifying certain patterns of inference.

Although this choice of units and population is standard in the time series analysis, Sober's claim of probabilistic dependence between SL and BP is obviously based on another choice, which is explicitly stated by Steel (2003). Steel writes: "if we take the units of the population to be time intervals, say one year each, then V [SL] and B [BP] are probabilistically dependent." (Steel 2003, p. 310) Relative to this choice of units, the population, as Steel points out, is a mixture of populations with different joint distributions over the two variables SL and BP. Based on a rehearsal of an old result on mixing populations — which Steel calls Mixing Theorem[e] — Steel shows that in the mixed population, the two random variables SL

[c]There are several notions of stationarity in the time series literature. The two most useful notions are *covariance stationarity*, which roughly means that the variance and covariances (between, say, X_t and X_{t+s}) do not change over time; and *distribution stationarity*, which roughly means that the (joint) distributions of the variables (say, $X_t, X_{t+1}, \cdots, X_{t+s}$) do not change over time. A non-stationary time series $\{X_t\}$ is *integrated* (of order 1) if the first-order difference $\{X_t - X_{t-1}\}$ is (covariance) stationary. Two integrated time series $\{X_t\}$ and $\{Y_t\}$ are said to be *cointegrated* if there is a linear combination of the two series, $\{\alpha_0 + \alpha_1 X_t + \alpha_2 Y_t\}$, that is (covariance) stationary. The extension of the notion to higher-order and to multiple series is straightforward. See e.g., (Enders 2003) for the precise mathematical definitions.

[d]We are not completely convinced, though, that cointegration is a perfectly adequate surrogate for probabilistic dependence between different series as Hoover seems to suggest, but this is not relevant to the present issue.

[e]The mixing theorem says that given two distributions in each of which X and Y are probabilistically independent, that is, $P_i(X,Y) = P_i(X)P_i(Y)$ $(i = 1, 2)$, then mixing them (by whatever

and BP are indeed probabilistically dependent (i.e., $P(SL, BP) \neq P(SL)P(BP)$), assuming that SL and BP are independent in the subpopulations before mixing. So he concludes, contrary to Hoover's evaluation, that Sober's counterexample stands.

Given Steel's choice of units, we find no fault with his argument. Yet the argument leads to a conclusion opposite to Hoover's. We hope to have exposed that their disagreement lies with the choice of units; they are working with different populations and variables. There is no real conflict between their conclusions. The real issue is whose choice of units is more appropriate for statistical and causal analysis.

The diagram in Figure 1 presents an inferential perspective on the issue. The solid arrows in the diagram depict what we might call direct inference. In particular, properties of the probability distribution may be derived from features of the causal structure (based on principles like the CMC), and statistical properties of the sample can be inferred from the population distribution (together with assumptions about the sampling scheme). The dotted arrows, on the other hand, represent the inverse inference. Note that there are two big steps in the statistical inference to causality: first to infer features of the population probability distribution from sample statistics — statistical estimation or testing; and then to infer causalities from probabilities — causal inference (cf. Papineau 1994). Based on this picture our view can be summarized as follows: relative to Hoover's choice of units, there is no problem with the CMC, which figures in the second step, but Sober's sample statistics would not work for the first step; relative to Steel's choice of units, the problem lies in the second step even though the sample statistics indicate a real population property in the first step.

If we are right about the crux of the problem, it seems most cogent to ask the following question: which choice of units and variables is better for the sake of causal analysis in Sober's case? A comprehensive discussion of the relevant principles governing the choice of units is beyond the scope of this paper. We will instead highlight a principle particularly pertinent to the CMC and causal inference in general. The principle is that units (and usually variables as well) should be chosen in such a way as to internalize causal relations between (instantiations of) variables *within* units, rather than to keep them *across* units. A peculiar feature of the population chosen by Steel (and Sober) in the SL-BP case is that there is presumably causal interference across units relevant to the variables of interest. By contrast, there is no such inter-unit causation given Hoover's choice of units. This difference suggests yet another diagnosis of the situation, namely, that the causal Markov condition is not a safe assumption in the presence of inter-unit causation.[f] Let us elaborate on this point in the next two sections.

---

ratio) will give rise to probabilistic association between X and Y unless $P_1(X) = P_2(X)$ or $P_1(Y) = P_2(Y)$. This point can be traced back to Yule (1903) — who gave credits to Karl Pearson on this matter, and is cited by several other authors to comment on the SL-BP scenario (e.g., Meek and Glymour 1994; Spirtes et al. 2000).

[f]This diagnosis was briefly mentioned in Spirtes et al. (2000, p. 296).

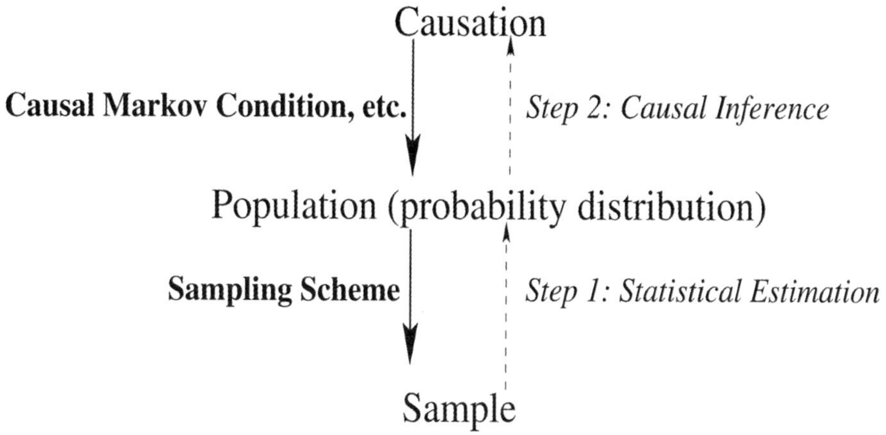

Fig. 1. A two-step approach to statistical inference of causation

3. Causal Models and Intra-unit Causation

The CMC is a principle that links a qualitative causal structure to patterns of probabilistic dependence and independence. Corresponding to different conceptions of probability there could be different versions of the CMC. In the context of statistical analysis, it seems most natural to formulate the CMC with respect to a population probability distribution, which is what concerns us in this paper.[g] The qualitative causal structure among variables can be represented in several equivalent ways. One standard representation that can best illustrate the point is the *functional causal models* (Pearl 2009, pp. 26-38), which generalize the linear structural equation models familiar to econometricians and social scientists. A functional causal model consists primarily of a triple $\langle X, \mathcal{E}, \mathcal{F}\rangle$. X is the given set of substantive variables, $\{x_1, ..., x_n\}$. $\mathcal{E}$ is a set of error terms, one for each variable, $\{\epsilon_1, ..., \epsilon_n\}$. $\mathcal{F}$ is a set of deterministic functions[h] $\{f_1, ..., f_n\}$ that satisfy a set of structural equations

$$x_i = f_i(PA(x_i), \epsilon_i), \quad i = 1, ..., n$$

where $PA(x_i) \subseteq X \backslash \{x_i\}$ denotes, intuitively, the set of parents or direct causes of x_i relative to X. Probability can then be introduced by stochastic laws on the error terms.

[g]This, as it sounds, seems to lean towards a frequentist conception of probability. A Bayesian version of the CMC is formulated, for example, in (Meek and Glymour 1994), which becomes a normative principle governing degrees of belief relative to a known (or believed) causal structure. For friends of propensities, they can formulate a single-case version of the CMC, say, with respect to a single unit. Then it would be trivially true that the causal structure depicts intra-unit causation. We want to emphasize, however, that even when the CMC is formulated with respect to a population, the causal structure at issue does not include any inter-unit causation whatsoever.
[h]Steel (2005) argues that the deterministic functions can also be used to represent genuine indeterministic causal relations if we do not treat the error terms as causes.

In a functional causal model, each structural equation is interpreted as an autonomous mechanism that describes how the instantiations of some variables *in a unit* influence the instantiation of another variable *in the same unit*. The italicized phrase is the crucial point: the modeling scheme only represents causal relations within units. After the units are chosen, there is no representation of any causal relation across units unless we redefine what a unit is to internalize the said relation. For instance, if we define a unit to be a single person and choose a population of persons, including you and me, then the causal influence of my education level on my income is representable in the framework, but the causal influence of your education level on my education level or my income, if any, is not represented in the model. To be able to represent such influence we have to redefine a unit to be, say, a pair of persons.

This point appears to be commonplace among practitioners of causal inference using statistics. Note, however, that the reason why inter-unit causation is not represented is not that the concept of inter-unit causation is unintelligible. To repeat a point we made earlier, units are chosen on largely pragmatic grounds for the sake of analysis, and hence are extrinsic to causation. One may have a ontological notion of "Unit" as whatever entity that assumes the relata of an instance of causal relation. But such Units do not necessarily correspond to the units chosen in statistical analysis. The latter, as we described in the previous section, can be combined to form larger units or decomposed into several smaller ones for various purposes. Causal relations, by contrast, are real features of the world (or so we assume), which cannot be created or destroyed by simply choosing units in a certain way. One and the same instance of a causal relation could be either intra-unit or inter-unit, depending on what are chosen as units.

Then why does inter-unit causation fail to be recognized in the current formal language of causal modeling? This is, we think, due to an important characteristic of statistical inference, with which the causal models are designed to align. Although units are extrinsic to causation, they are intrinsic to sampling. Statistical methods rely on samples, which are samples from a population of units. Without specifying units beforehand, it does not make much sense to talk about samples. After the units and variables are laid down, statistical analysis is concerned solely with (the distribution of) the joint instantiations of the variables within units, but not across units. In that sense, statistical analysis is confined to dealing with intra-unit relationships.

This is not saying that statisticians cannot study the relationship between samples. On the contrary, statisticians talk about independent versus correlated sampling all the time. But such discourse makes sense only because the (implicitly chosen) unit is enlarged. For example, to say a sample of size n, $\langle x_1, ..., x_n \rangle$, for a random variable X is, for example, i.i.d (identically and independently distributed) is actually to treat each data point x_i as an instantiation of a distinct variable X_i. Thus not each single sample point but rather the whole vector $\langle x_1, ..., x_n \rangle$ is treated as a unit. Therefore, to say samples are independent or correlated is not to invoke any inter-unit relationship. But the remark is indeed relative to a different choice of units and population than the original one from which the samples are drawn. There is surely nothing novel in this observation, which simply exemplifies the common practice of choosing different units for different analytical purposes.

Accordingly, statistical modeling of causation aims at intra-unit relationship only. Any assumption that provides a bridge from probability to causality, such as the causal Markov condition, is formulated in terms of intra-unit causal relations. It is thus important that units should be chosen in such a way to avoid relevant inter-unit causation, the presence of which is potentially detrimental to the task of inferring intra-unit causation.

4. Inter-unit Causation and Causal Inference

To appreciate this methodological advice, let us explore in more detail how inter-unit causation might frustrate causal inference. For a clear illustration of how inter-unit causation endangers the CMC, consider the two scenarios in Figure 2 where a sharp contrast is set between a population with inter-unit causation and a population without. The arrows in the figures indicate causal influence. Suppose, in both Figure 2(a) and Figure 2(b), the variables are chosen to be X and Y (among possibly other variables), and the units of the population are chosen to be $U[i]$'s that instantiate X and Y to $X[i]$ and $Y[i]$ (among possibly other variables). This choice of variables and units is obviously appropriate for the situation depicted by Figure 2(a) in that causal relations are within units rather than across units. For Figure 2(b), however, there is causal interference across units. To make the case even simpler and clearer, suppose both X and Y are binary, and the mechanisms in (b) are: $X[1] = Y[1] = 0$, $X[n+1] = 1 - X[n], Y[n+1] = 1 - Y[n]$. It is then easy to calculate that in the (infinite) population X and Y are probabilistically dependent: $P(Y = 0|X = 0) = 1 \neq P(Y = 0) = 1/2$.[i] But there is no (intra-unit) causal connection between X and Y, and hence the CMC is violated.

Clearly this example is designed to mimic the causal structure in the case of Venetian sea levels and British bread prices. Suppose both the sea levels and the bread prices form integrated time series, then a reasonable picture of what is going on is the structure in Figure 2(b). One might try to explain the failure of the causal Markov condition in the example by remarking that it is a case of mixture of two populations and it happens that the partition variable (i.e., time) is metaphysically unsuitable for being a cause. But this diagnosis misses a prominent causal feature underlying the mixture. It is the inter-unit causation that propagates an initial coincidence into a genuine probabilistic association.

In another article, Steel (2005) uses the following example due to Berkovitz (2002) in order to argue, contrary to what some people believe, that determinism

[i]Interested readers can easily come up with parameters for Figure (a) so that its population has the exact same joint distribution over X and Y. The same population distribution, however, provides information about causal connection in one case, but supplies misinformation in the other.

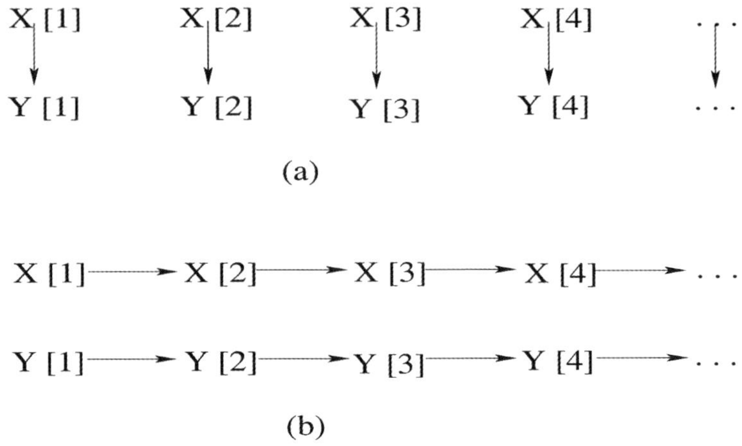

Fig. 2. Intra-unit causation vs. Inter-unit causation

does not give more warrant to the CMC than indeterminism. The example presents two "slot machines constructed entirely independently of one another but which, coincidentally enough, have precisely the same initial conditions and internal mechanics." (This is of course reminiscent of the famous two clocks in the history of philosophy.) Steel's point is that if we take the outcomes of the machines as variables of interest (and world-at-a-time as units), the two variables will be correlated (and hence violate the CMC) even though the machines are deterministic. However, it is transparent that this case of parallelism has the causal structure depicted in Figure 2(b), and so violates the CMC due to the operation of inter-unit causation.

The presence of inter-unit causation may also affect the first step of causal inference, the statistical inference from sample to population — as described in Figure 1 — by making independent sampling harder. Consider the following twist to Figure 2(b). Take the chain to be a time series with the exact same degenerate transition probabilities as before, but suppose that the initial probability distribution is this: $P(X[1] = 0) = P(Y[1] = 0) = 1/2$ and $P(X[1]Y[1]) = P(X[1])P(Y[1])$ (i.e., $X[1]$ and $Y[1]$ are independent). It is not hard to check that this series is stationary according to the usual definition, and hence $X[i]$ is independent of $Y[i]$ for each i. Again, following Sober and Steel, let us take the variables to be X and Y, and the units to be worlds-at-a-time. The population for X and Y is a combination of populations for $X[i]$ and $Y[i]$. Moreover, since the series is stationary, in the combined population X and Y are still probabilistically independent. However, based on a single realization of the series, no matter how long it extends, calculating the sample correlation as Sober does for sea levels and bread prices will always give 1 (or -1). So the sample correlation based on a single realization — dependent sampling — of the series would be very misleading.

In general this is not the case for well-behaving stationary time series. The peculiarity here is due to the extreme transition probabilities, given which the dependence of late samples on early samples does not diminish as the distance between them increases. Still, it serves to illustrate how dependent sampling, a close relative to inter-unit causation, might bias the statistical inference of population properties (and more often decrease the efficiency of statistical inference). Although the presence of interference between units renders correlated sampling more likely, it is important not to conflate inter-unit causation with dependent sampling and their respective consequences. For one thing, independent sampling is still possible given a population with inter-unit causal influence. It is just that the natural sampling schemes that would result in independent samples were there no inter-unit causation may produce correlated samples instead. More importantly, inter-unit causation and dependent sampling in general figure in different stages of causal inference. Back to Figure 1, dependent sampling is an issue in the first step, whereas inter-unit causation is primarily an issue in the second.

The potential failure of the causal Markov condition is not the only worry in the presence of inter-unit causation. A well known assumption in the experimental and observational studies of treatment effects is the *stable unit treatment value assumption* (SUTVA) (see, e.g., Rubin 1986), which rules out the possibility of inter-unit interference. The rationale behind the assumption is quite transparent. If inter-unit interference is present, for example, in the study of a treatment on a infectious disease, it will not be possible to consistently estimate the average (intra-unit) treatment effect. In the extreme case, inter-unit causation may propagate a treatment effect to the whole population so that there is no correlation between treatment and recovery, even though there is indeed an intra-unit treatment effect. This would be a violation of what is known as the *causal Faithfulness (or stability) condition* (Spirtes et al. 2000; Pearl 2009). There is also a debate over the status of the causal Faithfulness condition, but neither the critics nor the defenders seem to have pointed out that the condition is more shaky when there is inter-unit interference than otherwise.

It has thus been fairly common in practice to choose units so that the relevant inter-unit causation is avoided as much as possible. For example, when the SUTVA is believed to be violated, say, in the context of a study of an epidemic, the units are usually enlarged from single persons to groups of people, and the treatments on different people in a unit-group are regarded as distinct variables (cf. Halloran and Struchiner 1995). So the principle that we wish to highlight in this paper is actually observed in practice. Philosophical discussions, we believe, can greatly benefit from attention to such principles.

5. Conclusion

Sober's Venice-Britain counterexample to the causal Markov condition, in our view, is based on an improper choice of units. Given that choice of units, the variable SL (Venetian sea level) and the variable BP (British bread price) are indeed probabilistically dependent in the population, as Steel correctly shows. However, given the more appropriate choice of units, the process SL and the process BP are not probabilistically dependent, and the sample correlation calculated by Sober cannot be used to judge association in the population, as Hoover rightly points out. Since a change of units usually involves a change of variables, our analysis is in line with the general theme that for the causal Markov condition to hold, one has to choose the "right" variables (Hausman and Woodward 1999).

Our analysis highlights the methodological principle that units should be chosen in such a way to avoid inter-unit causation relevant to the chosen variables. One may object to this principle on the ground that it appears to be epistemically circular: to apply the principle requires some causal knowledge in the first place. We have two responses to this objection. First, the principle is indeed at work in many practical situations, where practitioners do not seem to have much trouble identifying the proper units. In particular, it is often independently testable whether there is causal interference across the chosen units. Second, it is not necessarily a disaster that inference of certain causal relations or quantities has to rely on some background causal knowledge. We feel no qualm about conceding that "no causes in, no causes out." As long as the "causes in" are not the same as the "causes out", no question is begged. It is not unreasonable to expect that in at least some situations where a choice of units is explicitly called upon, there is adequate information to guide the choice. Given longitudinal data, for example, it is usually advisable to conform to the standard choice of units in the time series analysis. The problem of such a choice is that there is often only one realization of the series, i.e., only one data point. But then background knowledge or assumptions such as stationarity and/or invariance of causal structure may come to the rescue.

In view of the difference between Steel's and Hoover's choices of population, some readers might have been prompted to draw the following seemingly sensible moral: the causal Markov condition is particularly problematic with respect to non-homogenous population, and one should avoid applying the Markov condition to a mixture of populations with different probability distributions. This proposal, however, appears inadequate for at least the following reason. In practice, almost every population one encounters can be partitioned into sub-populations with different probability distributions over a set of variables. The causal Markov condition would become practically vacuous if it did not apply to any mixture of populations. In fact, the phenomenon that combining different populations may create new probabilistic associations is seldom regarded as a serious threat to the causal Markov condition *per se*, but rather as an illustration of the importance of causal sufficiency for the causal Markov condition (Spirtes et al. 2000, p. 34, Hausman and Woodward 1999, p. 528). It seems to us that the most we can conclude from Sober's case is that *some* cases of mixing populations do pose a special problem for the causal Markov condition, but there is no reason to think that all cases do. An interesting project is to figure out the conditions under which mixing populations creates a problem other than violating causal sufficiency. This paper provides one candidate for such conditions.

References

1. Berkovitz, J. (2002). On causal inference in determinism and indeterminism, in *Between Chance and Choice: Interdisciplinary Perspectives on Determinism*, Eds. H. Atmanspacher and R. Bishop. Charlottesville, VA: Imprint Academic, pp. 237-78.
2. Enders, W. (2003). *Applied Econometric Time Series*, 2nd ed. Danvers, MA: John Wiley.
3. Hausman, D. M. and J. Woodward (1999). Independence, invariance and the causal Markov condition. *The British Journal for the Philosophy of Science* **50**: 521-83.
4. Holloran M.E. and C. J. Struchiner (1995). Causal inference in infectious diseases. *Epidemiology* **6**(2):142-51.
5. Hoover, K. D. (2003). Nonstationary time series, cointegration, and the principle of the common cause. *The British Journal for the Philosophy of Science* **54**: 527-551.
6. Mill, J.S. (1843). *Systems of Logic*, vol.1. London: John Parker.
7. Meek, C. and C. Glymour (1994). Conditioning and intervening. *The British Journal for the Philosophy of Science* **45**: 1001-21.
8. Papineau, D. (1992). Can we reduce causal direction to probabilities?, in *PSA Procceding*, Eds. D. Hull and K. Okruhlik, Vol.2, East Lansing, MI: PSA, pp. 238-52.
9. Papineau, D. (1994). The virtues of randomization. *The British Journal for the Philosophy of Science* **45**: 437-50.
10. Pearl, J. (2009). *Causality: Models, Reasoning, and Inference* (2nd ed.). Cambridge, UK: Cambridge University Press.
11. Reichenbach, H. (1956). *The Direction of Time*. Berkeley: University of California Press.
12. Rubin, D. B. (1986). Which Ifs Have Causal Answers?. *Journal of the American Statistical Association* **81**: 961-2.
13. Sober, E. (1987). The principle of the common cause, in *Probability and Causation: Essays in Honor of Wesley Salmon*, Ed. J. Fetzer, Dordrecht:Redel, pp. 211-28.
14. Spirtes, P., C. Glymour, and R. Scheines (2000). *Causation, Prediction and Search* (2nd ed.) Cambridge, MA: MIT Press.
15. Steel, D. (2003). Making time stand still: a response to Sober's counter-example to the principle of the common cause. *The British Journal for the Philosophy of Science* **54**: 309-17.
16. Steel, D. (2005). Indeterminism and the causal Markov condition. *The British Journal for the Philosophy of Science* **56**: 3-26.
17. Yule, G. U. (1903). Notes on the theory of association of attributes in statistics. *Biometrika* **2**: 121-34.
18. Yule, G. U. (1926). Why do we sometimes get nonsense-correlations between time-series? A study in sampling and the nature of time-series. *Journal of Royal Statistical Society* **89**: 1-63.